MIKE BARA

THE CHOICE

Using Conscious Thought and Physics of the Mind to Reshape the World

New Page Books
A division of The Career Press, Inc.
Pompton Plains, NJ

Copyright © 2011 by Mike Bara

All rights reserved under the Pan-American and International Copyright Conventions. This book may not be reproduced, in whole or in part, in any form or by any means electronic or mechanical, including photocopying, recording, or by any information storage and retrieval system now known or hereafter invented, without written permission from the publisher, The Career Press.

The Choice

Edited by Kate Henches

Typeset by Eileen Munson

Cover design by Howard Grossman/12E Designs
Printed in the U.S.A. by Courier

To order this title, please call toll-free 1-800-CAREER-1 (NJ and Canada: 201-848-0310) to order using VISA or MasterCard, or for further information on books from Career Press.

The Career Press, Inc.
220 West Parkway, Unit 12
Pompton Plains, NJ 07444
www.careerpress.com
www.newpagebooks.com

Library of Congress Cataloging-in-Publication Data

CIP Data Available Upon Request.

For Alyssa Reid Leblanc,

for teaching me absolute forgiveness
of who we have been,
absolute appreciation of who we are,
and for giving me absolute excitement
at who we can become.

If only we so Choose....

Acknowledgments

I would like to acknowledge the following people for their inspiration and their support in this project, and my life in general.

My brother Dave; Alan Pezutto; my sister Kelli; Barbara Yates; Neena Dolwani; Susan Shumsky; William Henry; Jay Weidner; my mom Millie; MC Burton; Hollywood Meoil; the fetching Dr. Mignon Walker-Marquina; Chandra Marie, who taught me how to love again; Indy, my guide in all things; Cleo, who I can't wait to see again; and my little light Aurora, who showed me the value of unconditional love.

Special thanks to my friend and mentor, Richard C. Hoagland, without whom this book would most likely not exist.

And my deepest thanks for the Gift of Denise Zak, without whom I might not have ever reached this amazing place in my life.

Contents

Introduction 9

Chapter 1: The God Force 15

Chapter 2: The Original Solar System 31

Chapter 3: Hyper-Dimensional Physics 45

Chapter 4: The "Laws" of Physics 59

Chapter 5: The Physics of Time 71

Chapter 6: The Golden Alignment 81

Chapter 7: The Science of the Cycles 91

Chapter 8: The Mayan Calendar 101

Chapter 9: The Mayan Factor 113

Chapter 10: The Hopi Prophecies 123

Chapter 11: Torsion 133

Chapter 12: The Explorer Effect 143

Chapter 13: The 2012 Venus Transit 159
Chapter 14: The Nemesis Effect 173
Chapter 15: The Doomsday Scenarios. 189
Chapter 16: Signs and Wonders 209
Chapter 17: The Choice 219
Chapter Notes. 227
Selected Bibliography. 231
Index. 233
About the Author 239

Introduction

This planet is in trouble.

Most of us know that. Or at least we sense it in some indescribable way. There is an air of unease all around us. A tone in the background that implies that all is not as it seems, and that what little stability we feel beneath us is subject to being shattered and torn asunder at a moment's notice.

We should feel this way. There is every good reason to. But it may not be for the reasons you think.

Invariably, when the news media talks of dangers to planet Earth, they focus on various "crises" that seem to fade rather quickly into the woodwork. Global warming, the housing crisis, the homeless crisis, the medical insurance crisis, they all come and go. And recently, we seem to be in constant cycle of crises; one after another after another until it seems like we are replaying some twisted scene out of *1984*. Yet, contrary to what you see reported on the 24-hour news channels, the human condition keeps improving generation after generation thanks to our collective wisdom, innovation, and compassion for our fellow man. Every year, infant mortality declines, we cure more diseases, we feed and house more human beings worldwide, and our collective IQs go up.

So if the truth is that most of our worries are really nothing at all to worry about, why are we so restless? Why do we all feel so upset and concerned for our futures, both in the short and long runs? Why are we so anxious for change, for a new way of living? The answer again is fairly straightforward if somewhat unbelievable: It's because the stars are making us feel this way.

The reality is that many of our most sacredly held beliefs and support systems *are* on the brink of being ripped from us completely. That all this is happening on the precipice of another stressful event, the end of the current Mayan calendar long count cycle, will only add to the sense of dislocation most of us are already feeling. After all the conflict, political polarization and financial turmoil of the last decade, we now have to face *this*? The potential end of all things at the hands of some irresistible Force that will consume us in…what? Hellish solar flares? Horrific earthquakes on a scale we have never experienced? Worldwide torrents of rain caused by a magnetic pole flip? Or the ultimate catastrophe—a geographic pole shift—as depicted in the Hollywood movie *2012?*

Or will it instead be, as so many Mayan scholars and New Age leaders insist, a momentous change not in our physical existence, but in our *consciousness*? Will it be a new awakening that will finally end the Occult age in which we now live and allow us to move forward in truth and harmony with our God?

Truthfully, the transformation we all feel coming may be none of those things. Or it may be all of them. The only thing we know for sure, deep in our souls and in our bones, is that we *are* going to see a change. The only question is what form that change will take.

Believe it or not, that Choice is entirely up to us.

Let's be honest, the idea of a spiritual transformation of our consciousness is more comforting to most of us. It implies that, although we may not have quite so much material abundance in our lives, we at least will have food to eat, shelter, and perhaps even a renewed sense of community. It evokes an image of living a simpler, quieter, timeless life, one where we can focus on the spiritual development of our hearts rather than the materialistic development of our minds. As so many New Age gurus tell us, the Mayan calendar is a road map to the evolution of human awareness, and nothing more.

But what most in the New Age community don't realize is that there is an underlying physics, not related to Einstein or quantum mechanics or gravity, driving these consciousness shifts. We'll get into the details of this "hyper-dimensional" physics a bit later, but make no mistake about it: An energetic pulse is coming from the center of our galaxy sometime in the next several years. The change we are going to experience, whether in our conscious minds or our physical reality, is influenced by this pulse.

The ancients, who lived in a time when we were much closer to our Maker than we are today, understood this at a fundamental, almost instinctive level. They saw the dimension of time as a great circle, a series of cycles that repeated endlessly and gave us the chance to revisit our past mistakes, and to learn from them and atone for them. As they looked up into the skies and counted the relentless motion of the stars, they saw that this particular period in our history—the post millennial era—was one of great significance. The Egyptians marked it by the rising and falling of Orion's belt when viewed from the Great Pyramid on the Giza plateau. The Hindu Vedas saw it as the culmination of a series of astronomical cycles of the Earth orbiting around the source of all life, which they called the Vishnu Nabhi. The Mayans marked it by the passage of Venus across the Sun and the once-every-26,000-years alignment of our solar system with the center of the Milky Way galaxy. The Hopi Indians of North America marked it with a series of prophecies of which only one or two remain to be fulfilled. The Christians spoke of it in their Book of Revelation with the passage *Revelation 20:12*.

And each of these traditions tells of the time in which this transition will occur in blunt, sometimes frightening language. They speak of a cleansing of the Earth, of wars between the good and the wicked, of Choices to be made, sides to taken, and battles to be fought. What I will show you about this is twofold. First, all of this, even the great cleansings, are only possibilities driven by this little understood physics. There is a long history of suppressed and ignored experimentation that shows that this physics can not only be moderated, it can be manipulated and controlled. It allows us, through our own thoughts, prayers, and true desires, to dictate the outcome of this next decade of change.

Second, I believe the truth of what the ancients knew has been lost in the distant translations of time, and we have far more control over what will happen in the coming 2012 era than the doomsayers believe. There were originally at least 1,562 Mayan Codices. Only four survive today. For all we know, the other 1,558 could tell in great detail of the various ways in which we will grow and learn and all walk off into the sunset singing *Kumbaya* in 2012. Perhaps it is only the famed and feared Dresden Codex that warns of a chaotic future time. What I will make a case for in this book is that all of these scary prophecies, the cleansings, the great shifts, the battles, are going to be fought not in the physical, but in our hearts and minds and spirits at the unconscious level. It is there that we will take our stands and make our Choices, and it is here, in our 3D reality, that we will make those Choices manifest.

But that is by no means guaranteed. We have to Choose our path. And soon.

We have been deluded into believing that the Apocalypse is something that happens *to* us, rather than understanding that it is something we actively participate in. What happens when the world as we know it comes to an end could be traumatic. Or it could be merely *dramatic*. It's up to us.

The energy pulse itself is neutral; it has no inherent intent (good or bad) until it manifests in our 3D reality and tunes to the frequency of our planet. If we are a world enmeshed in fear, greed, and conflict, the results could be significantly—if not catastrophically—negative. But if our world is tuned to a higher tone, seeking harmony, love, and peace, then these energies can be channeled into a positive force, and the dire predictions of the Maya's Dresden Codex, the Hopi prophecies and all the rest can be lessened by a factor of hundreds, if not thousands.

In order for this to happen, as many people as possible must be made aware of the potential perils ahead. They must be given the tools to do their part in creating a better world, in joining with God in co-creation, and in setting the tone for the next phase of human development: the Universal Consciousness. We can do this by creating a second Harmonic Convergence, a grassroots movement much like the original Convergence in 1987, that will attract its participants with a positive—but serious and urgent—message to humanity. And unlike the first one, this Harmonic Convergence must be maintained. It must last far longer than a single weekend and become a completely new state of *being*.

As you will see in these following pages, there really is no argument between the various interpretations of the Mayan calendar or any of the other prophetic readings of this period in our history. They all agree that a great Shift is coming. But what we have not understood until now perhaps is that it's the *physics* that drives the *consciousness*; it's the *science* that drives the *spirit*; it is the *astronomy* that drives the *astrology*. We need to accept that there is a science behind our thoughts, moods, and actions, and that means we can counteract any negative energy that they generate in the midst of our coming existential crisis.

The occult age of Pisces is ending. The new age of Aquarius is dawning. No one can stop it any more than they can stop the precession of the Equinoxes. The "Darwinian delusion"—that we are nothing more than a random collection of biological accidents that evolved in a creation without a God—has poisoned our souls for far too long. That's what the ancient Maya and the Egyptians and the Vedas understood. Now, *we* get to write the next chapter in the book of our own destiny.

All we have to do is look deep into our hearts and our souls.

We've all learned The Secret. We each have the power to create our own reality through our hopes, dreams, or, yes, even our fears. Armed with this alchemical knowledge, it is now time for us to put that power to use, to make The Choice.

The time has come.

Who are you? And what do you want?

1
The God Force

Right up front, I want to make clear to you what my premise is. The physics that we are taught in school—Einstein, Newton, et. al.—is wrong. Quantum physics, the bastard stepchild of Einstein, is also wrong. There is no such thing as dark matter. There is no such thing as zero-point energy, super strings, or maybe even the Big Bang. From the beginning, our modern understanding of physics has been hampered by the fact that we have been restricted to an incomplete model, a mistaken belief that we live in a purely three-dimensional universe that is closed to outside (higher) influences. As a consequence, modern astronomers and cosmologists (those that study the origins of the universe) are forced to move the goalposts to keep their broken theories alive when new experiments undercut their predictions.

What we have been missing, as Richard C. Hoagland, my co-author of *Dark Mission—The Secret History of NASA*, once said, is that Newton and Einstein aren't the whole picture. They are only a subset of a much richer fabric of reality that *must* include not only a fourth, but a fifth, a sixth, and all the way up to a 26th dimension at the least. Only when you go back to James Clerk Maxwell (1831–1879) and his original ideas for harmonizing electro-magnetism and gravity do you have a chance to step back, pick up the pieces, and see the whole picture. From that vantage point, it quickly becomes obvious that our universe is *hyper-dimensional*, consisting of ever-higher levels of existence, each of them a bit closer to God himself.

In order to unify electro-magnetism, Maxwell had to rely on something he called the "Aether," which he perceived as a fluid, wave-like field that existed everywhere, between everything, and that connected us all with everything else (sounds kind of like "the Force" again, doesn't it?). This Aether was the very necessary intervening medium between everything that could conduct the electrical or magnetic forces from one place to another. But as he worked on the equations, he quickly ran into the same problems as later workers like Einstein and even the string theory guys: He couldn't get there from here. For reasons too complicated to go into here (although they are explained in detail in *Dark Mission*), Maxwell realized that what he needed to complete his equations was the introduction of something new to the math: a fourth spatial dimension.

Once he did that, he found that the math worked itself out just fine. As he saw the universe, anyone could manifest any amount of energy at any time and in any amount (electrical potential) at any point in the known universe. It didn't have to come *from* anywhere because it all came from somewhere else anyway—the fourth dimension. To fix his equations, he used special numbers, called "quaternions," that worked only in the fourth dimension. But sadly, Maxwell died before he could complete the entire body of his work and publish it. The work was then carried on by another man named Oliver Heaviside. Heaviside was a contemporary (and intellectual rival) of Maxwell's, and he detested the hyper-dimensional aspects of Maxwell's work. He felt that the use of the quaternions was "mystical, and should be murdered from the theory." And when Maxwell died in 1879, that's exactly was Heaviside did.

The end result is that all of the classical electromagnetic equations that are attributed to Maxwell are actually *Heaviside's* equations. And, because they lack the crucial hyper-dimensional component—the fourth spatial dimension—modern physics was forever locked into basing all subsequent theory on an incomplete thought, a subset of a much richer notion of the universe that can explain everything we see, touch, and feel…including our thoughts, emotions, and our consciousness itself.

In our brief but richly physical existence here on Earth, we can only touch, see, smell, and hear what takes place in this three-dimensional realm. But we can *feel* those higher dimensions, *taste* a bit of what they must be like, and experience our connection to them through energy that flows into our universe through the stars, and into our consciousness through our hearts. And all of this experience is mechanized, like some giant celestial clock, through the geometry of our solar system.

That is not to say we don't have free will. We certainly do. We always have a Choice. But what I will show you about this physics is that it does set up the circumstances from which we will Choose.

God does this through the alignments of the planets in space around us. As you'll see in coming chapters, there is a large experimental database that proves that common everyday energies (things like radio waves) are actually affected by the positions of the other planets in our solar system. Their specific geometries, whether Saturn or Jupiter are at a 90-degree position to our own Earth's orbit for instance, have been shown to dramatically affect the quality of shortwave radio signals here on Earth. What this means is that if radio waves can be influenced by the positions of the planets, then our own thoughts, moods, and dreams can be affected, too. After all, what are our thoughts except brain waves, which, like radio waves, are simply another form of electromagnetic energy? What *that* means is that astrology, long dismissed by modern science as a silly superstition, is *real*. And not only that, as I will show you in a bit, all of this electromagnetic energy comes from somewhere else, some place higher than this simple three dimensional universe. And *that* implies not only that we each have an immortal soul, one that is in constant contact with us through our thoughts, feelings, and actions, but then it follows that there must be a God, too. After all, somebody had to set this all up, right?

See where this is going?

In case it has escaped your notice, I'm not going to shy away from the use of the word God in this book. Call it what you will: the Universe, the collective unconscious, the Universal Consciousness, the Force, the Higher Power, the Great Spirit, it all means the same thing. We're all talking about the all knowing, all loving, and all giving heart that gave birth to everything. I'm just going to cut through the crap and call it God.

Our God has given us a wonderful jewel in the night upon which to live out our lives. We call it the Earth. At one time, there were at least two other worlds suitable for advanced life in this solar system. Sadly, they did not survive to the present day and right now Venus and Mars are all but uninhabitable, as least as far as human life is concerned. As I look around this region of space we call home, I see, much like our own world, that it is one messed-up place. But it is also a place of intense beauty and geometric perfection. And as we take a brief tour of it in subsequent pages, you'll see how perfectly it is all connected, how everything in it makes it possible for us to inhabit our mother Earth.

Scientists and skeptics, who have been trained since an early age to reject the magical and to worship the hyper-rational, will look at the things I will point out to you in the coming pages and scoff. They will ridicule you for not realizing that it is all just due to mathematics and gravity and magnetism, along with various other naturally occurring phenomena. "Naturally," at least by their definition, means without the involvement of a God-creator. In short—it's just the way the numbers came out. Where I split from them, the "walking dead" as Robert Temple describes them in *The Sirius Mystery*, is that I look at those numbers, at those forces, and I see perfection. Not the perfection of random, meaningless existence for existence's sake, but the perfection of a grand, living, and breathing force of love that unifies all of us, from the biggest, densest black hole down to the simplest bacteria. Yes, there is a Force out there, and its name is God.

For generations, an argument has ranged back and forth in science and in theology as to the nature of the God force. There are various references to it in all major and minor religions, and it goes by many names. The Hindus and Buddhas call it *Prana* (vital life), the Chinese call it *Qi* (life energy), and in the New Age spiritual religions we generally refer to it as the *Aether*. In the western Holy Bible, we pretty much ignore all that and just call it God. But that also raises a question: Just who, or what, is "God"? As my friend Sean David Morton likes to say, "God" equals G.O.D.: Generator, Operator, Destroyer, the maker and destroyer of all things. But of course, God has many other names.

The Elohim

The name of God appears in the Bible some 4,473 times in 3,893 different verses. But the vast majority of these references are to a specific entity named "*Yahweh*," also known as *Jehova*, or YHWH. Anyone who has read Zachariah Sitchin's *12th Planet* series will know that Yahweh is most likely not a god of any kind, but rather a visitor from some other place and time. But there is also a second reference to God in the bible. It takes the form of the masculine plural word *Elohim*, which is used singularly only 30 times—all in the first chapter of *Genesis*—and it has a different meaning. It refers to a single, all knowing, all loving God force that exists everywhere. *Yahweh Elohim*, the Lord God, only starts to appear in Genesis 2:4 and that reference seems to be to a master over the Earth or sub-servant of the one true God, *Elohim*. Thus, the Lord God is not a god at all, but rather a representative of God here on Earth. *Elohim*, despite its plural origin, seems to have an entirely different interpretation. *Elohim* is the unifier, the whole, the merging of East and West, masculine and feminine, rational and spiritual. More simply, *Elohim*, God, Prana, Qi—whatever you want to call it—is...

One.

So when science and spirit go in search of God, it is not Yahweh or Allah or Great Spirit or anything or anyone specific that is being sought. It is the One. The Unity. The Love. The life force.

Of course, in our left brained, rational, intellect dominant Western world, such an elegant and self-evident concept as that is something that few of us take seriously. Only those brave enough to admit a commitment to the spirit, those who spend their time in quiet thought rather the hyperactive pursuit of the material, have any real proof of the existence of the God force. And that proof is all in our hearts, thoughts, feelings, and experiences. And we all know that the Western world places very little value on such evidence.

But the truth is that there really is plenty of hard and fast evidence of the *Elohim* around us. Even the most shut-down and closed-minded intellectual skeptic would have to admit such evidence is compelling, assuming they are open to looking at it. No matter how hard the hyper-rationalists argue that the God force is something that can never be touched, that we can never know it in our minds the same way as we know it in our hearts, they simply cannot deny the evidence that will presented in this book. The *Elohim*, despite its ethereal nature, is something we have been documenting in science for decades, if not centuries.

One of the first hints of a true God force that could be observed measured or touched was in the form of something called Kirlian photography. A Russian inventor named Semyon Kirlian discovered quite by accident in 1939 that when a photographic plate is placed within a high-voltage electrical field, anything on the plate will leave a distinctive colored halo or "corona" around its perimeter. Skeptics, as they always do, attack the notion that the halos are any kind of reflection of the living force of life energy, arguing that they are simply caused by moisture around the edges of the objects being imaged. Not one of these skeptics has ever conducted an experiment that *proved* this assertion, by the way, but nevertheless it is still made on a fairly regular basis.

One of the big problems with the moisture argument is the vast knowledge base of experiments done by various researchers and amateurs with living plant life. It was Kirlian himself who conducted the first "leaf experiment," where he placed a leaf he had just cut from a plant and photographed it quickly using his electrified plate process. The result was a beautiful image of a shimmering electrical halo around the leaf.

But when he removed the leaf, cut it, placed it back on the plate, and took another image, the severed portion of the leaf still remained, although it was weaker and more faded than it had been in the first image. The skeptical argument is that the faded corona is simply caused by the evaporating moisture from the severed portion left on the plant, but again this overlooks the fact that the leaves are rarely if ever placed back on the plate for the second image in exactly the same place as they were for the first, yet the faded corona is always aligned perfectly with the now severed portion of the leaf.

Other, later researchers discovered similar—but not identical—phenomena. One experiment found that when a leaf (which they only *intended* to tear into two pieces manually) was placed on a Kirlian photographic plate, the portion that they had decided to tear off did not fully develop. To their surprise, the undeveloped portion exactly matched the line of the tear, even though the tearing was done by hand *after* the first image had been taken. It was almost as if the Universe could see a few moments into the future and knew exactly where and how the leaf would be torn.

Speculation as to what exactly Kirlian photography is capturing ranges from the skeptical claims of simple moisture droplet patterns to what the New Age community calls the "aura," an energetic field surrounding all living things that connects us to the God force. In the 1970s and 80s, studies were conducted at both UCLA and the California Acupuncture College Clinic that showed a substantial increase in the intensity of the Kirlian photography results after acupuncture treatments, especially around the traditional Chinese energy points. Acupuncture, which has been practiced at least as far back as the second century Before Christ, involves the placement of tiny metal needles in certain "energy points" along the body. These needles are thought to clear the neural pathways so the life energy (the "Qi" or "chi") can flow more freely along the body and promote healing, mood alteration, and spiritual well-being. As before, the skeptics are quick to attack the results, although how the Kirlian photography could have known the location of the traditional acupuncture energy points and made them somehow more moist than the other parts of the bodies that were photographed is a mystery.

Other experiments involved male-female couples. One experiment at the Heuristic Institute in California showed that when the couples were neutral or mad at each other, the coronas generated in the Kirlian photographs stayed separate or didn't appear at all. When they sent thoughts of love toward each other or kissed, the coronas merged, elegantly reflecting the feelings of closeness and unity that the couples must have been feeling.

All of these experiments are fine, as far as they go, but although those of us that are spiritually awake instantly appreciate their significance, others, in a sleepier mood, will find little to sway them from their firm academic indifference. The next set of examples is harder to ignore.

In 1966, a polygraph (lie detector) expert named Cleve Backster decided to conduct a series of experiments on plants and other living things he had in his lab. Backster was the director of the Keeler

Polygraph Institute and had previously worked for the CIA in their interrogations unit. At one point, he decided to hook up a common houseplant he had in the lab to the polygraph. After leaving the plant hooked up for some time to get a baseline of its electrical activity, he began to consider some other ideas to measure any changes in the electrical output of the plant. None of these were very fruitful, until he had the thought that it might be interesting to see what burning one of the plants leaves might do to the readings. To his surprise, the instant that he had the thought, the polygraph went wild, jumping up and down as if the plant was in mortal terror of what he was about to do. Intrigued, he then began a series of experiments with other plants and found that, whenever he had similar thoughts, the plants that were hooked up would have similar reactions. Beyond that, Backster's research found that distance and even electromagnetic shielding were no impediment to the results.

Even more bizarrely, Backster found that not only was the effect non-local, meaning that the plants reacted even if the person initiating the thought was hundreds of miles away, but it was also *instantaneous*. Now, according to Einstein and relativity, both ideas are impossible. You simply cannot have instantaneous action at a distance in three-dimensional space. It just can't happen. According to the "laws of physics," nothing—not radio waves, energy, or thoughts—can travel faster than the established speed of light, which is about 186,000 miles per second. Yet in experiments conducted with former NASA astronaut Dr. Brian O'Leary, Backster proved exactly that. From more than 350 miles away, O'Leary sent out the intent to torch some of his own blood samples had given to Backster, and the cells—hooked up to a lie detector machine—responded instantly and with no light time delay. Beyond that, Backster eventually found that the plants could read his actual *intent*. In other words, they could tell the difference between a sincere thought of doing harm to the plant versus simply stating it internally without any real intention of following through with the act.

After years of conducting research on everything from brine shrimp to human blood cells, Backster found that all living things seemed to be connected by some powerful force that was present all around us at all times. In the past, this force was called the "Aether," a sort of fluid energy field that exists everywhere. Today, we might call it the Force.

I think it is very important to distinguish the difference between the Aether, or *Elohim*, as I've described it, and an aura. The Aether would be the all connecting energy field that exists everywhere at all times, and is the backbone for all of creation. It's the ocean of life energy through which the thought signals in Backster's experiments passed. An aura is something quite different. It's an intense life energy field that surrounds a specific entity—like you, for instance—and it is through this aura that we are all connected to the Aether itself, as well as to each other.

The Princeton Global Consciousness Project

One of the most stunning proofs of the existence of this God force comes from something called the Global Consciousness Project, or GCP. Created by a loose consortium of scientists, lay researchers, and simply curious individuals, the GCP was founded in 1997 and sought to use the communications capabilities of the Internet to determine if in fact there was a global consciousness tying us all together. Led primarily by Dr. Roger Nelson of Princeton University's Princeton Engineering Anomalies Research lab, the original vision was simply to find a way to measure consciousness and intention in the physical world. The first experiments used 12 specially placed hardware random number generators (technically called Random Event Generators, or REGs), to measure the effects on random number generation when human thought was specifically focused on certain events. The REGs act essentially as automated coin flipping machines, and in the end, the data should show no discernable pattern.

In an attempt to get a baseline of data, the group used scripted events to focus consciousness. These ranged from psychotherapy sessions to worldwide public events like the World Cup of soccer, the Academy Awards broadcasts, and the reaction to the death of Princess Diana. The GCP even organized an Internet-based guided meditation called the "Gaiamind Meditation" in January 1997. What all this research showed was that focused human thought *did* seem to have a measureable effect on the random number generators, to the point that it was decided that a larger study was justified.

At this point, the Global Consciousness Project was expanded to create a global network of about 65 REGs placed in locations all over the globe. Connected by the Internet, these new generators pumped out thousands of "coin flips" every second. Using this system, the project was quickly able to create a database of randomly generated numbers. Very soon after this worldwide network went live, it underwent its first trial by fire.

A few hours before the September 11, 2001, terrorist attacks on the World Trade Centers in New York City, the REGs started to register an uptick in the random number generators. By the time of the actual attacks, the REG's were registering an even greater non-random distribution, and as the day wore on and the towers fell, as more and more people the world over began to hear of the attacks, the curve just kept climbing upward, peaking on the 13th of September 2001. Obviously, we all remember those first few tense days after the attacks, waiting for the other shoe to fall and not knowing if the enemy had other plans in mind, up to and including using pirated nuclear weapons in a U.S. city. By several days after the attacks, the curve still had not returned to normal, and the number generation remained dramatically higher than normal. A paper by Nelson, Radin, Shoup, and Bancel in *Foundations of Physics Letters* put the odds at hundreds to one against chance that the data registered was non-random. Or to put it another way, the numbers had *meaning*. The reaction of the computerized devices had been real, and it had been affected by our collective consciousness.

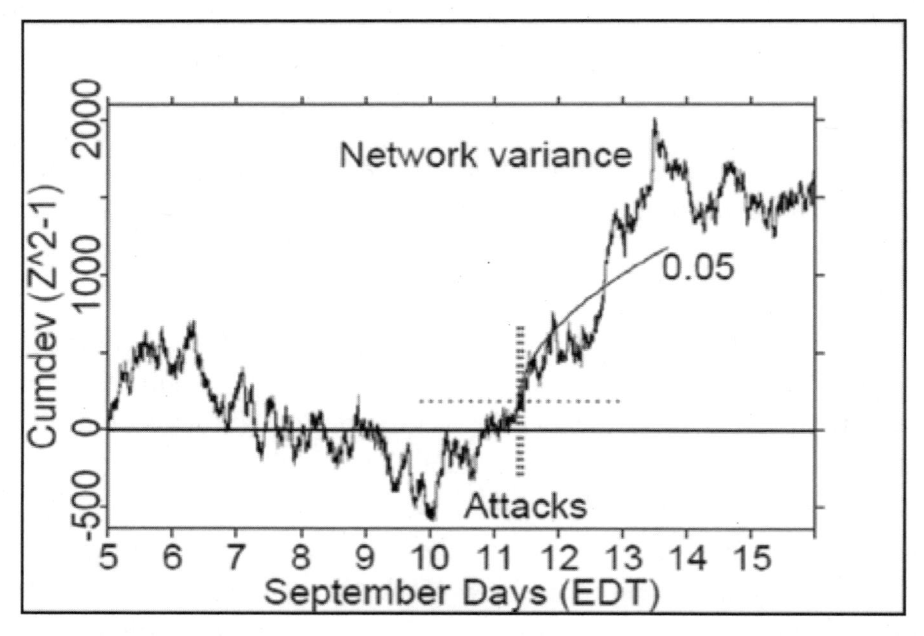

Image 1.1. GCP Data chart showing the spike in activity starting a few hours before the 9/11 attacks. Image courtesy of Roger Nelson and the Global Consciousness Project. Used with permission.

It was clear that the GCP had registered the proverbial "tremor in the Force," and that made it vital for the skeptics and critics to attack the data. They came on very aggressively, up to and including making sarcastic comments on blogs and constantly changing the Wikipedia page associated with the Global Consciousness Project to attack the results. Forced to defend themselves, the resident members of the GCP responded with papers of their own that showed that the data curves were unaffected by the early data, and beyond that showed that the early data was statistically valid in any case. At the time of this writing, the Global Consciousness Project has run tests around 324 global events, ranging from the 9/11 attacks, sporting events, the deaths of famous people like the Pope, and global days of peace. The results are plain and inarguable: To the tune of a *billion* to one, these number generators react when the consciousness of the world is focused on something significant.

It's easy to see why the skeptics feel compelled to attack the Global Consciousness Project. There really isn't anything that could be more threatening to their world view than proof of the existence of the Force, the Aether, the *Elohim*—call it what you will. But it is also just as easy to cut through their attacks and expose them as completely illogical as they are. Consider this: The whole premise of their academic attacks is not that the random number generators don't show unusually high readings during events that have a worldwide impact. They freely acknowledge that they do. What they seek to do instead is engage in an intellectual game called reductionism, which consists of finding some fault with individual data points—say the attacks of September 11, 2001—and then using that to claim that the overall premise is false. But let's take a closer look at what they are really saying.

Whatever their arguments about individual cases, there is no question that the REGs jumped on September 11, 2001, even beginning a few hours before, as if the Universe itself could sense a great evil about to be unleashed. The critics argue that the data are flawed, and therefore we can't trust it, or for that matter any of the other 324 global events they registered. But what if the Global Consciousness Project REGs hadn't registered anything at all on September 11th? Do you have any doubts that these same critics wouldn't now be citing *that* as evidence that the whole idea of an interconnecting Force among all living things was nonsense? Of course they would. Even if all of the other global events registered off the scale, they'd be citing the September 11th attacks as proof that the Force couldn't exist, because how could the ultimate global event of the decade, the defining moment of the century so far, not be part of the fabric of data backing it up?

The simple truth is, whether the event is expected (like the Olympics) or seemingly random like the terrorist attacks of 9/11, they always register on the GCP's computers. *Always*. And the results are always way beyond any possibility of data errors or random fluctuations. So my conclusion is pretty straightforward. There is a Force, and, though we may not be able to touch it in some specific way, we can certainly measure its effects on our physical world, or at least on our electronic instruments, like the random number generators.

The Flynn Effect

There is one other conscious phenomenon that is taking place right now and I believe it is somehow connected to the wave of change that is coming and to this God force all around us. There is great deal of evidence that certain types of human ability, specifically reflected on IQ scores, have been steadily rising beyond the expected rate of growth for as much as a century.

James Flynn is an American political scientist (now living in New Zealand) who has been studying IQ tests and their various results for decades. Originally, Flynn sought to prove that any racially distinct IQ results were because of cultural biases in the tests, not to any innate differences in the abilities of different racial groups. But what he found instead was that IQs in general were increasing at an almost alarming rate, and that they were doing so across all racial, cultural, and educational impediments. What became known as the Flynn Effect was intensely controversial from the time it was first brought to light in the 1980s, but subsequent tests have not only confirmed it, they have greatly deepened the mystery of its origins. In one Spanish study, the "mean IQ" (the middle range of all the children tested) had increased by 9.7 percent in 30 years. In other tests, researchers found increases across single generations of anywhere from five to 25 points. In other words, teenage children today are as much as *25 percent smarter* than their parents. Without adjustments made to "normalize" (that is, fudge) the data, more than half the children tested in 1930 would be considered to be bordering on mental retardation today. Worldwide, the phenomenon is not only prevalent, it is accelerating.

Traditional science has no explanation for this. By their standards, the increases are unaccounted for even when factoring in longer school days, the widespread use of computers, and video game use, which is thought to enhance certain types of learning. Flynn himself is reluctant to embrace his own data and argues that there is no evidence that people are actually getting smarter. If there was, he says, we'd be living in a second Renaissance, with art science and all forms of inquiry leading to new discoveries almost daily.

However, the truth is that there is a perfectly reasonable and obvious explanation (first proposed by Richard Hoagland) that simply hasn't been considered. Most IQ tests are multiple-choice. Students are usually given between two and four options to answer any given question, and are encouraged to take their best guess if they don't actually know the answer. So to me, the explanation is as plain as the nose on the Face on Mars.

It's not intelligence that's on the rapid rise; it is *intuition*. Plain, good-old human gut instinct. This also neatly explains why some people, like spiritual mediums, seem to possess skills that other "normal" humans do not—they are based on intuition. The heart instead of the mind. The soul instead of the personality. The spiritual instead of the material. What this means is that we could indeed be moving into a new, higher way of living, feeling, and interacting with each other. What academia don't seem to know is exactly what is causing it. To me, the cause is again completely obvious.

The stars are making it happen.

2
The Original Solar System

The notion that the stars or planets somehow affect our moods is as old as human civilization. As far back as the third millennium Before Christ, there are records of recognizable astrological charts, and it has only been in recent times that astronomy and astrology have been separated. Up until the last few hundred years, they were considered two sides of the same coin. There simply was no dispute that the movements of the planets had an impact on human events. Today however, it is obviously a different story.

As Wikipedia (for what it's worth) puts it:

> Astrology is a group of systems, traditions, and beliefs which hold that the relative positions of celestial bodies and related details can provide information about personality, human affairs, and other terrestrial matters. Astrologers believe that the movements and positions of celestial bodies either directly influence life on Earth or correspond to events experienced on a human scale.

Skeptics and rationalists, like the National Science Foundation, describe astrology as a "pseudoscientific belief"—something that only the great unwashed and uneducated masses believe in. The truth is, astrology is a perfectly valid and defensible science. And, the case to be made for the significance of astrology actually comes from the territory of the science geeks themselves: from the fields of astronomy and physics.

Astrology argues on its face that we are all connected—that the Earth, its people, the planets and the stars are all part of a giant mechanism that is driving and influencing us at key junctures in our lives. Astonishingly, science and physics have shown plenty of experimental results that argue exactly the same thing—they just haven't been recognized yet by the scientific community.

One of the things that most people, even in the scientific disciplines, either don't appreciate or don't realize about our solar system is that it *is* a system. All of the planets and moons are part of a larger, interconnected structure that extends far beyond the boundaries of our own Solar system to the center of our galaxy and even beyond. The Sun and all the planets that it gave birth to (we'll get to that in a moment) are linked, each relying on the other for heat, resources, and life-energy. Likewise, Jupiter and its nine moons would be another system all to themselves, a sub-system of our larger solar system.

It is impossible for me to look around at all of this synchronized perfection and come to the conclusion that we and our world are somehow nothing especially interesting or important. There are many tantalizing clues that lead me to think that this place, our home, is special. My belief is that this is a "designer solar system" that was specifically created so that higher life might live and grow here. One such clue is the fact that the Moon governs how fast the Earth spins, which is key to it being able to support human life. Without the Moon's calming influence, the Earth would spin so fast that the centrifugal force would most likely flatten us all like pancakes. The Moon also regulates and agitates the Earth's protective magnetic field, which is so crucial to protecting us from high intensity solar flares and other types of deadly interstellar radiation. The Moon also just happens to be exactly the right size and exactly the right distance from the Earth to create perfect solar eclipses, a phenomenon where the disk of the Moon perfectly blots out the disk of the Sun. The diameter of the Moon also turns out to be 2,160 miles at the equator, which is, as Richard Hoagland, my coauthor on *Dark*

Mission, might say, "an interesting number." It's not 2,161 miles, or 2,159 miles across, but exactly 2,160 miles. And those 2,160 miles also just happen to match the 2,160-year length of each astrological age of the precession of the equinoxes. Some people think this is all just a coincidence—that the Moon just happened to be ejected from the Earth by some random planetary collision or somehow wandered into Earth's gravitational influence and stabilized in this perfect position from which it can protect us, regulate us, and even influence the female menstrual cycle.

All of this matters because, as I stated previously this is a true solar system. Beyond that, it is a *harmonic system*, one that was set to a specific frequency for a specific purpose. All harmonic systems are by definition in one of two conditions: resonance or dissonance. If a harmonic system is in *resonance,* it means that all parts of the system are able to transfer energy and information at the same frequency and with maximum efficiency, or *harmony*. If a harmonic system is in *dissonance,* that means that there is a conflict between some or all of the parts of the system, and all the parts must be brought back into resonance in order to restore harmony. The dissonant parts of the system also tend to align to the frequency of the larger whole to try and get back into that vibrational harmony.

The problem we face today is that our solar system is messed up to great extent, meaning that it is in dissonance right now. We can easily tell this just from a little poking around our local area. All of the planets orbit the Sun basically around its equator, something called the "plane of the ecliptic" by astronomers. All of the planets also spin on their own axes, and these are supposed to be perpendicular, or 90 degrees to the plane of the ecliptic. (See Image 2.1.) But the Earth is tilted off its proper vertical spin-axis alignment by some 23 degrees. Uranus is spinning literally on its side, at 90 degrees to its proper vertical spin axis alignment. Venus even rotates *backward*. Many of the planet's orbits, which due to the effects of the Sun's gravity, should be perfectly circular by now (4.7 billion years after the birth of the solar system), are highly elliptical (shaped like an egg). In fact, Mars's orbit is so "eccentric" (the astronomer's term for it) that its distance from Earth goes from 34 million miles at its closest to 249 million miles at its greatest.

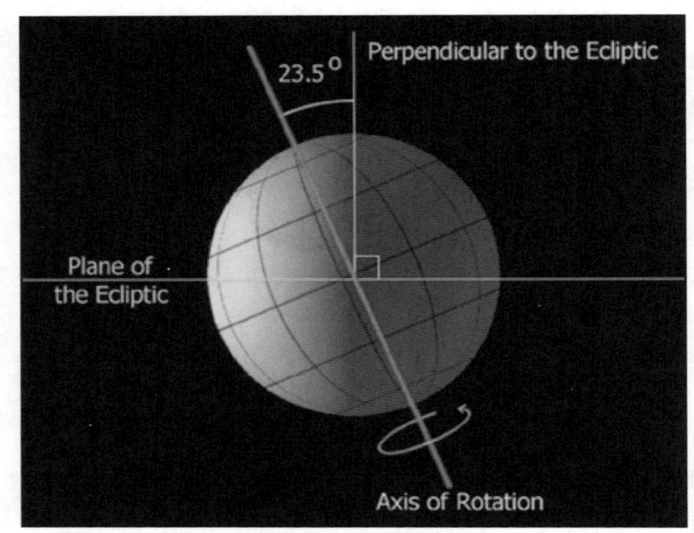

Image 2.1. The Earth spins on a rotational axis that is 23.5 degrees off of true vertical to the Sun's equator, accounting for the change of seasons, among other effects. Image by Krys Lilly.

So when I conclude that our solar system is messed up, what I mean is that there is dissonance in the system itself, for reasons I will get into a bit later. More importantly, what we must do is get this planet back into resonance with the rest of the greater solar system. And the best way to do that, in my opinion, is to understand just how and why the system got messed up in the first place, so we can understand what we can do to fix it.

The Birth of the Planets

According to current theories, planets, like our Earth, form from the dusty remnants of stellar nebulas created when a dying star explodes in a super nova. The "nebular hypothesis" (which has never actually been observed) argues that planets form in so-called "proto-planetary disks" when tiny grains of dust from these stellar nebulas begin to clump together because of electrostatic charges and a process known as accretion. After a while, the accretion of the nebular material forms a clump between 1 and 6 miles in diameter. Eventually, these clumps of material begin to run into each other, forming much larger clumps called planetesimals. This process then continues until the planetesimals get big enough to become what are classified as minor planets. At that point however, the models begin to break down, because the number of collisions drops off considerably as the growing minor planets sweep the stellar nebula clean, like massive cosmic vacuum cleaners. The theory then requires a series of coincidental planetary collisions, which, instead of breaking both objects into smaller pieces again, somehow (according to the theory) manage to force and compress them into ever larger balls of material, which then officially become planets.

All well and good. Except it's wrong.

The bottom line is that there are all sorts of logical problems and obvious contradictions with the accretion model. The biggest problem—besides it never having been observed—is its built-in complexity. Not only does it require a series of unlikely planetary collisions as its base presumption, it also assumes, contrary to common sense, that these cataclysmic collisions (at thousands of miles per hour) would in the end be *constructive*, rather than *destructive*, to the embryonic planets trying desperately to become big enough to take their place in the solar system. Beyond that, if all the planets started forming at the same time (estimated to be 4.7 billion years ago), they should all be about the same size. They clearly aren't. This also creates another problem. It is only possible for one planet to "eat" large chunks of another after a collision if one of the planets is substantially bigger than the other one. Bodies of equal size and speed would more likely obliterate each other completely.

Fortunately there is another theory of planetary formation that is supported by *all* of the evidence, called the "solar fission" theory of planetary formation. Some scientists, like the late Dr. Tom van

Flandern, have long argued in favor of it as the only viable model of how planets truly form. In his book, *Dark Matter, Missing Planets and New Comets: Paradoxes Resolved, Origins Illuminated*, Dr. van Flandern argued persuasively for the fission theory.

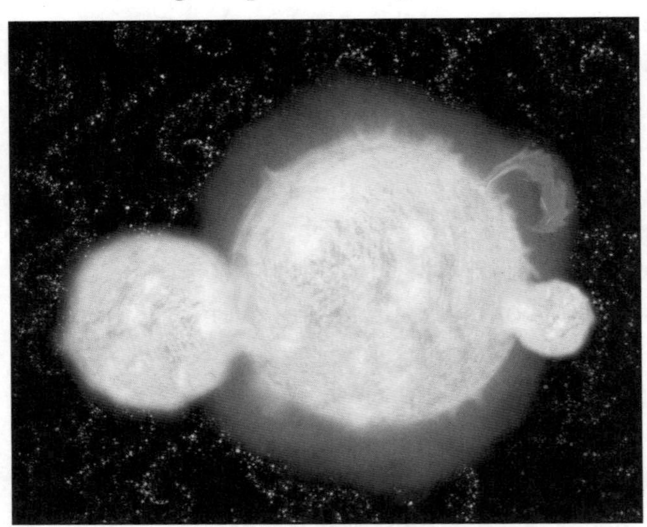

Image 2.2. Illustration of the solar fission theory of planetary formation. Image by Krys Lilly.

In the solar fission model, once the biggest chunk of the solar nebula collapses and begins nuclear fusion (ignition into a star), it starts sucking up all the nearby dust and gas and it quickly grows in size. By adding all this fuel to its nuclear furnace, it soon begins spinning so fast that the centrifugal forces become stronger than the gravitational field of the newborn star. At this point, the star "oblates" or bulges at the center, and solar material is flung out from the equatorial region of the young star. This material then spirals outward in (roughly) twin pairs, forming first gas giant planets and then later "terrestrial" or rocky planets like our Earth. As the star gives birth to pair after pair of twins in this manner, its angular momentum (spin energy) dissipates and the star begins to emit energy in a stable cycle ideal for supporting life giving planets. As the blobs of ejected "star stuff" spiral away from their birth mother, they also give birth in turn to their own moons, and eventually find their resonant orbits and begin to cool. After a billion years or so, the whole system should achieve a state of equilibrium and balance. The planets will cool. On some of them in the habitable zone, like Venus, Earth, or Mars, water will form oceans, bacteria will start the cycle of life, and the children of this elegant birthing process will eventually walk the face of these planets, stare into the night sky, and wonder how they got there in the first place.

This theory, unlike the solar accretion theory, neatly fits most, if not all, the observational data we have about planets, including the ones we see around other stars. Recent observations of the first extrasolar planets—planets that have been found orbiting other relatively nearby stars—show strong evidence that van Flandern's solar fission theory is correct. From the start, these observations (beginning in 1998) were baffling to the astronomers. What they found were large, massive gas giant planets like Jupiter orbiting almost impossibly close to their parent stars. In some cases, they would have been within the orbit of Mercury had they been in our own solar system. Not only that, but they were almost impossibly hot—in some cases almost as hot as the stars they orbited. The accretion model theorists didn't have a clue how this data could be real, and they were initially so flummoxed by it that they suspected their data must have been wrong.

In fact, it is the *accretion model* that's wrong. What we are observing is the early stages of a gas giant planet that has just been ejected from its parent star and is still in the process of slowly spiraling away and cooling. In the solar fission theory, these "Hot Jupiters" make total sense, whereas the accretion model must be completely rethought just to accommodate these "Hot Jupiters."

The solar fission theory would also explain a host of otherwise-unexplainable mysteries of the solar system. Why, for instance, are all the planets in the so-called plane of the ecliptic, the imaginary plane around the equator of the Sun? Why do the planets hold 98 percent of the total solar system angular momentum (spin energy), but just a fraction (0.002) of the mass? Why do all of the planets have prograde (counter-clock wise) orbits matching the prograde spin of the sun? Why did the planets in our own solar system form in relatively similar pairs (Earth and Venus, Mercury and the Moon, Jupiter and Saturn, Uranus and Neptune)? In the accretion model, none of these are predicted. In the solar fission model, they are *required* for the theory to be correct.

Beyond that, there is a certain harmonic grace to the solar fission theory that the accretion model just doesn't possess. What could be more poetic, more appropriate, and more *elegant* than the idea that the Sun gave birth to the planets, in the same natural way that mammals give birth to babies? In the solar fission model, all the planets are more than just a connected mechanical/mathematical system; they are a family, with the Sun as their mother in the same way that the Earth is our mother.

So the idea is that the Sun once possessed all of the spin energy in the entire solar system, and it just got going so fast that it started spewing out pieces of itself from around its belly button. These hot clumps of stellar material then spiraled outward, collecting more debris as they did and probably enhancing their size along the way until they settled in their current stable orbits. According to van Flandern's math, the larger outer gas giants (Uranus, Neptune, Saturn, and Jupiter) would then spin off smaller moons in the same manner until they, too, had little mini solar systems orbiting around them. These little moons would also form in twin pairs, spinning away from the bigger planets until they too stabilized and cooled, and became the rocky satellites we see today around Jupiter and Saturn. Equally, if Dr. van Flandern is right, the inner and rockier "terrestrial planets" like Earth and Venus would tend to excrete one large moon, as opposed to a bunch of little ones.

Okay, all is well and good. Except…

Well, for starters, we seem to have missed one planet completely. Ignoring the still raging debate over Pluto being demoted from planetary status to being a mere "dwarf planet," the fission theory as I've laid it out ignores a really important planet here that doesn't seem to fit into any of this; Mars. Not only that, but what about Mercury, the closest planet to the Sun? And if Venus fits the pattern, where the heck did its big moon go? Surprisingly, not too far at all.

The Exploded Planet Hypothesis

Dr. van Flandern's solar fission theory has always faced objections from mainstream scientists, most notably around the "twin pair" concept. Arguments are raised pointing out that Venus lacks a moon of any kind, that Jupiter is more massive than Saturn (in the twin pairs model it should be the other way around), and that Mars does not fit this model at all. However, Dr. van Flandern has another even more revolutionary idea in which all of this is easily explained, and not by some fairy-tale, move-the-goalposts band aid, but by actual, observational evidence. It's called the Exploded Planet Hypothesis.

The EPH, as we'll call it for short, really got its start way back in the latter half of the 18th century. Astronomers at the time noticed that all of the (then) known planets (Mercury, Venus, Earth, Mars, Jupiter, and Saturn) followed a regular spacing pattern. The Titius-Bode Law of Planetary Spacing, or simply "Bode's law," as it came to be known, observed that the planets all seemed to be about twice as far from

Titus-Bode Law of Planetary Spacing

Planet	Distance	Formula
Mercury	0.4	0.5
Venus	0.7	0.7
Earth	1.0	1.0
Mars	1.5	1.6
?	—	2.8
Jupiter	5.2	5.2
Saturn	9.5	10.0
Uranus	19.2	19.6
Neptune	30.1	38.8

Formula: Distance in au = 0.4 + 0.3 * 2(n – 2)

Image 2.3. Chart showing the spacing of planetary orbits according to Bode's predictions.

the Sun as the previous planet. Using this simple rule as a guideline, astronomers soon noticed a problem: There was huge gap between Mars and Jupiter where the fifth planet should be.

This led to telescopic observations of the gap and the eventual discovery of Ceres, at first thought to be the missing fifth planet but later determined to be merely a "dwarf planet," like Pluto.[1] However, the search for the missing "Planet V" did lead to another major discovery: The gap was filled with thousands of small rocky bodies called asteroids. The region where Planet V should be was then renamed simply "the asteroid belt." It didn't take much after that for Heinrich Wilhelm Matthäus Olbers (1758–1840) to suggest that the asteroid belt contained the remnants of an exploded planet, which he named "Phaeton."

But the gradualists, those who believe change happens very slowly through time, were having none of that. They moved quickly to shoot down Olbers's ideas, and within a few years both Olbers and Bode's law were on their way out of fashion. After the discovery of Uranus and Neptune, which didn't fit Bode's formula as well as most of the inner planets, the gradualists pretty much shot down any discussion of Phaeton or Bode in the scientific journals. That didn't stop van Flandern, who made new arguments in favor of both Bode and Olbers in his 1998 book.

As he pointed out in *Dark Matter, Missing Planets and New Comets,* the arguments against Bode, especially, were weak. If the solar fission theory were correct, then such a law of planetary spacing, with its two-to-one orbital resonance pattern, is exactly what we would expect to see. Each successive planet that was ejected from the Sun in this manner would rob the Sun of some of its own spin energy, again neatly explaining why the planets have so much of the spin energy in the solar system compared to the Sun. The first planets would obviously be the farthest out, spin the fastest, have the most satellites, and be the biggest. Logically, they would also follow the "twin pairs" concept, with the first body ejected being the biggest and the second body being somewhat smaller. As we simply look at the planets, we can verify that what we see fits this pattern really pretty well (excluding the dwarf planets like Ceres and Pluto).

Going from the outside in, we see that Neptune is indeed slightly larger than Uranus, as we would expect, but Saturn is slightly smaller than Jupiter, which does not fit the pattern. Going in from there, we see the asteroid belt, followed by Mars. Because we really have no idea if there was in fact a planet in the orbit of the asteroid belt and therefore have no idea how big it was, comparing it to Mars seems impossible. Next in, we see Earth and Venus, which fit the pattern precisely, Earth being larger than Venus.

So at first glance, it seems we are 50/50, at best. Earth and Venus and Uranus and Neptune fit the pattern, but Saturn and Jupiter and Mars and the asteroid belt do not seem to fit. Mercury and Pluto don't even get mentioned. So what gives?

It's all a matter of what you define as a "planet," for one thing. According to all the conventional models, there are eight planets in our solar system: Mercury, Venus, Earth, Mars, Jupiter, Saturn, and Uranus. Ceres and Pluto are mere dwarfs, and as such don't count. What van Flandern says is that there were originally 12 planets in our solar system, of which six remain today. I agree that there were once 12, but I also say there are eight still remaining. In van Flandern's model, the fission pattern would alternate: two big planets in twin pairs, followed by two smaller planets in twin pairs. As you initially look at the solar system, that seems to be wrong, at least on the face of it. But as you go deeper, it all adds up.

We all look at Jupiter and Saturn and assume they are the biggest objects in the solar system. But if there were two other planets that got spun out first—two planets we didn't know about and have never observed—then this changes everything. In this case, you'd have two undiscovered planets way beyond the orbit of Neptune (about 7.25 billion miles out and 14.5 billion miles out respectively) and two missing terrestrial planets in the orbits of the asteroid belt and about where the orbit of Mars is now. This obviously has big implications, not the least of which is that Mars isn't a planet at all, but perhaps merely a moon of a much larger and more massive planet that doesn't exist anymore.

So the original solar system must have looked something like this from the farthest out in toward the Sun: the two undiscovered planets, which I will call Nemesis and Nibiru, Neptune and Uranus, Saturn and Jupiter, the two destroyed planets, which I will call Phaeton and Maldek (after the Aetherius Society's name for it), and the two inner worlds, Earth and Venus. In my model, Mercury is an escaped moon of Venus, similar in size and composition to our own Moon, Mars is the sole surviving moon of Maldek, and Ceres the sole surviving moon of Phaeton, which is evidenced only by the remnants of it in the asteroid belt. Pluto also drops out of the equation because it is an escaped moon of Neptune, and nothing more.

This fits the big/small pattern perfectly, and also fits Bode's law almost to a "T." But if this was in fact the original configuration of our solar system, then what the hell happened to it? How could two planets, one near Mars's current orbit and one between Mars and Jupiter, have simply up and disappeared? The answer is pretty dramatic and somewhat frightening. It appears that Olbers's was right all along: They exploded.

Van Flandern argued that there was lots of evidence to support the idea that Maldek and Phaeton exploded. For one thing, so-called long period comets (comets that have never been observed before) come from all angles and directions, exactly as they should if they were in fact debris from the explosion of one of the two missing terrestrial planets. Conventional wisdom is that comets come from the same place, the mythical "oort cloud" beyond Neptune. But if that's truly the case, then they should all be in the plane of the ecliptic, where all the planets are, because they would merely be leftover remnants from the original clumps of rock that made up the early solar system. Only if there was an explosive event would comets enter into the inner solar system at such a wide variety of angles.

There is a great deal of other evidence to cite in support of the idea that Maldek and Phaeton exploded at some point in the past, but we may as well cut through the muck and get right to it. All we have to do is look at Mars.

The evidence that Mars is not a planet at all but a moon of a much larger world is written all over the face of it. As Richard C. Hoagland and I described in *Dark Mission—The Secret History of NASA* (Chapter 9), Mars shows all kinds of signatures that it was once a warm, habitable world devastated by a horrific explosive event that ripped away half the atmosphere and killed the planet in a single day. Besides the fact that Mars has what is known as a crustal dichotomy, a condition where the crust on the lower hemisphere is (in places) 20 miles higher than in the northern hemisphere, the most compelling evidence I can cite is Richard's simple observation that Mars has tidal bulges, two massive geologic uplifts 180 degrees apart. This is a flat-out guarantee that Mars was once in a tidal locked condition with Maldek, always showing the same face to her exactly as our own Moon does to us today. Maldek was probably what scientists are now calling a Super Earth, a rocky terrestrial planet twice as large and about five times as massive as our own Earth, but with far more tectonic activity (earthquakes) and far less stability.

Probably orbiting no more than 25,000 miles away from Maldek (judging by the size of enormous Tharsis and Arabia bulges on Mars), Mars was once a blue-green paradise, much like Earth, with vast oceans, rivers, and valleys, and who knows how many different kinds of life. As Richard and I argued in *Dark Mission*, I believe there is ample evidence that Mars was also host to a very advanced civilization, but that is another issue altogether. In fact, I strongly suspect that Venus, Earth, and Mars were all once capable of supporting higher forms of life.

These planets all reside within the so-called "habitable zone" in our solar system, which coincides with the region roughly between the orbits of Venus and Mars. It is so named because, in this region, temperatures are warm enough to support the development of oxygen based atmospheres, liquid water, and planets with liquid cores. All of these ingredients are thought to be crucial to life here on Earth. As I see it, there was once an inner earth (Venus), an outer earth (Mars), and our own abode of life.

And ours is now the only one left.

What is important to understand is this: It is the number of planets and their spin energy that dictate the total overall energy of our solar system. If there are now missing planets in the form of Maldek and Phaeton, that means that not only is the physics of our solar system *different* than it was before they were destroyed, there is now less life energy in the system than there used to be. Richard argues that's why the dinosaurs no longer roam the Earth, and why we have no animals that even remotely approach their size on our planet any longer. The system is broken. Whatever happened, by whatever cause, has changed the physics of our solar system, because the physics is driven by the geometric configuration and the total energy in that system.

That's why Earth wobbles on its axis now, why that spin axis is inclined at 23 degrees, why our planet struggles to align itself to the new harmonic frequency of the greater whole. What we must do is try to find out just how to help her get back into resonance before she is forced to adjust on her own. But to do that, we have to understand just what really drives Life, the Universe, and Everything. And I'll give you a hint: It ain't quantum physics.

3 Hyper-Dimensional Physics

It is one thing to say that there is a new and better theory of physics than the ones we are taught in schools, and quite another to prove it. But after more than 15 years of looking at all of the various arguments, I'm now convinced there is a far more logical, organic, and harmonic "theory of everything" that works in all circumstances. As I will show you in Chapter 4, it can't be quantum physics, or relativistic physics, or any of the other ideas we are taught in the world's colleges and universities. The rules that the universe seems to adhere to best fall under the veil of something called *hyper-dimensional* physics.

I first read about the hyper-dimensional physics model of the universe in Richard C. Hoagland's book *The Monuments of Mars*, in the early 1990s. The name comes from the simple idea that everything we see and experience in this universe as energy actually originates from somewhere else—from outside our observable 3D realm. The word *hyperspace*, from which the term *hyper-dimensional physics* is derived, was first coined in 1854 by German mathematician Georg Riemann. He wrote the first mathematical science paper actually describing the geometry of a spatial dimension—the fourth—beyond this one. From then on, in the language of science anyway, higher dimensions were collectively lumped together and referred to as "hyperspace."

As I wrote with Richard in *Dark Mission*:

> The cornerstone of the hyper-dimensional model is the notion that higher spatial dimensions not only exist, but are also the underlying foundation upon which our entire 3D reality exists. Beyond that, everything in our observable 3D world is, in fact, driven by mathematically modeled "information transfer" from these higher dimensions. This information transfer might simply be the result of changes in the geometry of a connected system, say a change in the orbital parameters of a planet, like Jupiter or the Earth. Since we are limited in our perceptions to the 3D universe we live in, we cannot "see" these higher dimensions. However, we can see (and measure) changes in these higher dimensions that have a simultaneous effect on our reality. By definition, this change in higher dimensional geometry is perceived in our 3D universe as an "energy output"...

In other words, when planets and stars and even whole galaxies change their positions relative to each other, these actions create "a tremor in the Force" in higher, imperceptible dimensions. Because these resonant waves aren't restricted by our limited three-dimensional "laws" like the supposedly fixed speed of light, they can (and demonstrably do) have an effect in our universe at great distances—and instantaneously. This single notion (instantaneous action at a distance) in and of itself completely undercuts all our current models of reality, which are based on what you'll see are very inadequate "Laws of Physics" that scientists feel compelled to follow.

Let me see if I can find a simpler way to explain all this.

As kids, a lot of us did things like use a magnifying glass to heat up ants as they went about their business. Now, we'd never do that today, but kids aren't always so nice. The point is, what the ant is experiencing is heat energy from the Sun, amplified and focused by the magnifying glass being held over him. Because of his limited perception, the ant can't see the boy holding the glass, the magnifying glass itself, or even the Sun. And even if he did, there is no way that he could possibly understand what they were or how they were making him feel hotter. But, just like us in our three-dimensional reality, with its limited visual spectrum and limited sensations of sight, smell, and sound, the ant does experience the heat energy, and he can even measure it by walking around the edges of the area being warmed up. The same thing applies to us. Although we may not be able to perceive higher spatial dimensions or where the energy is coming from, we can measure the effects here on Earth, in our tidy little subset of the greater, hyper-dimensional reality.

In *The Monuments of Mars* and in an earlier paper entitled "*The Message of Cydonia*," Hoagland and his co-researcher, Erol Torun, argued that certain mathematical alignments of what appeared to be ancient ruins on the planet Mars (yep, I said it) seemed to be implying that there was an unknown or previously undiscovered Force operating in the universe. This Force left distinct signatures all over the solar system in the form of energetic upwellings found on almost all the major planets very near the latitude of 19 1/2 degrees. Neptune's Great Dark Spot, the Great Red Spot of Jupiter, the erupting volcanoes of Jupiter's moon Io, Olympus Mons on Mars (which is the largest shield volcano in the solar system), and Earth's own Maunakea volcano in Hawaii all were at, or very near, the 19.5° latitude. They later realized that the location of these energy upwellings was based on the geometry of a circumscribed tetrahedron, basically a four-sided pyramid surrounded by a sphere.[1]

This single numeric clue led Hoagland and Torun to the early research of Maxwell and also to a group of rogue mathematicians called topologists. They found that the topologists had done a great deal of theoretical research mapping the mathematical properties of a rotating "hypersphere," a sphere that exists in more than just our standard three spatial dimensions. The arcane math describing this "hypersphere" and the multiple dimensions (26) above it are so complex that they are virtually unintelligible to all but the most educated math buffs. But the math also said that the *signatures* of the higher dimensional states of this sphere would be observable here in its primitive, three-dimensional state. If the hypersphere was rotating (and remember: *Everything* in our universe is rotating at all times), then the math required that rotating roiling twisting motions, exactly like the observed rotational dynamics of Jupiter's Great Red Spot, would leave a marker on the sphere in 3D at a specific location: 19.5 degrees latitude.

In other words, if you take a sphere—like oh, I don't know, a *planet*, say—and you rotate it, then you will pull energy from the higher state of the planet (the hyper-sphere) and that energy will preferentially upwell from inside the planet according to the geometry of a tetrahedron encased in a sphere, and appear at or around 19.5 degrees. This is exactly what Hoagland and Torun observed all over the solar system. As they went even deeper into the math, it implied that the tetrahedron is even more than that; it's nothing less than the base building block of all the solid matter in the physical three-dimensional universe that we live in. In fact, tetrahedrons are used by video game programmers all over the world to create all the 3D objects in the games you play because they have found that this is the easiest way to build solid objects in 3D. There is also a certain kind of balance in the symbolic nature of the geometry of a tetrahedron circumscribed by a sphere, because the tetrahedron is the simplest of all solid geometric forms (the so-called Platonic solids) and the sphere is the most complex.

Let's take this even a bit further. If everything that is spinning (and remember: That's *everything,* literally) is putting out energy, then doesn't it follow that these energetic forces might in fact interact with each other in some way? Might these waves of energy generated by the spin of the planets have some kind of impact on our physical reality here on Earth? It's an interesting notion. If only we had some way to prove it, right? Well...

In the 1950s, RCA hired a young engineer named John Nelson in an effort to improve the reliability of short-wave radio communications around the Earth. Such radio transmissions were crucial to long-distance communications at the time because the high-frequency signals could be bounced off the Earth's ionosphere, a layer of electrically charged ("ionized") atoms that lies between the upper atmosphere and the Earth's magnetic field. RCA requested the study because they had noticed that the quality of the short-wave transmissions varied greatly depending on the Sun's sunspot cycles (it is solar radiation that "ionizes" the ionosphere in the first place). They had already found the short-wave signals to be more reliable in the lulls in between the solar activity associated with peak sunspot years.[2] The more active the Sun's sunspot cycle, the more the magnetic field of the Sun interfered with the short-wave radio transmissions.[3] Upon beginning his study, Nelson soon found that the radio interference rose and fell not only with the sunspot cycle, but also with the *motions of the major planets of the solar system.* He discovered that the relative positions of the so-called gas giant planets (Jupiter, Saturn, Uranus, and Neptune) along their orbital paths seemed to have the most dramatic effect on the signals:

> It is worthy of note that in 1948, when Jupiter and Saturn were spaced by 120°, and solar activity was at a maximum, radio signals averaged of far higher quality for the year than in 1951 with Jupiter and Saturn at 180° and a considerable decline in solar activity. In other words, the average quality curve of radio signals followed the cycle curve between Jupiter and Saturn rather than the sunspot curve....[4]

So what Nelson concluded is that while the sunspot activity of the Sun *did* have an effect on the quality of short-wave radio signals, the locations of the planets along their orbital paths (especially the major gas giant planets) had a *far greater effect* on the signals. Now, that may not really bother you too much, but the fact is there is nothing in any of the conventional mainstream physics theories that can account for this. Not gravity, not magnetism, nothing. The gravitational influence of these planets is negligible because of their great distance from the Earth and Sun. Although Jupiter, Saturn, Uranus, and Neptune are certainly massive, and we know that the more mass an object has the greater its gravitational field, they are so far away that their influence amounts to virtually nothing.[5] And though magnetic fields like the one the Sun generates demonstrably *do* have an influence on the Earth's ionosphere and hence the quality of the radio signals (which the RCA study proved), the magnetic fields of the distant gas giant planets are far too weak to have an effect, especially an effect that is actually greater than the Sun's much more powerful (and much closer) magnetic field. The only conclusion, then, is that somehow, someway, the planets were affecting the quality of radio wave signals on Earth from great distances, and by some means that was as yet undiscovered.

The questions that no one in the mainstream physics community could answer at the time (and they still can't) are exactly what that Force might be, and how it might propagate through the vacuum of space to create the sunspots and the subsequent magnetic activity. Once gravity and magnetism are eliminated, what's left? Well, pretty much nothing, at least in the conventional models—except electricity. We can rule out electricity though because it can't travel through the vacuum of space without a conductor like an electromagnetic plasma, which is a very rare phenomenon. So in the absence of any other plausible explanation, we are left with only one way the Force energy could have traveled the distance between the Sun and the outer planets and caused the increased magnetic storms (you guessed it)...

higher dimensions.

In the hyper-dimensional model of reality, such instantaneous interactions at distance are not only implied; they are built-in to the whole concept. If all you have to do to generate energy is to rotate something, then that means that everything is putting out energy: plants, trees, animals rocks, people—even (and especially) whole planets. As we look at these four gas giant planets (again, Jupiter, Saturn, Uranus, and Neptune) we find that they do indeed put out more energy than all the other planets combined. It's just not a kind of energy that modern physicists and astronomers recognize.

As we discussed earlier, the Sun holds 99 percent of the mass in the solar system. If the currently accepted "Laws of Physics" were correct, then that means that the Sun must also be the most influential body in the solar system, because it therefore has the most gravity and electro-magnetic energy. But if that assumption were correct, then Nelson's RCA experiments would have concluded that the Sun's magnetic activity had the greatest influence on the radio signals. Instead, it showed that the *planets* somehow had more influence, which

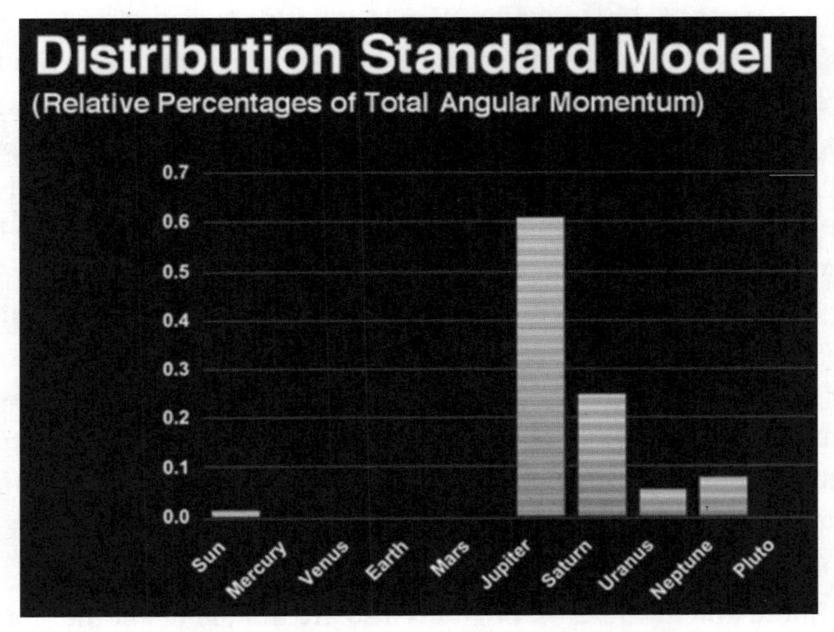

Image 3.1. Fully 98 percent of the total solar system angular momentum is held in the gas giant planets, Jupiter, Saturn, Uranus, and Neptune. Image by Krys Lilly.

is exactly what we would expect in the hyper-dimensional model of reality. Exactly how they do this is through something we've talked about before; a source of energy called "angular momentum," or spin energy.

Jupiter, as it turns out, has less than 1 percent of the mass in the solar system, yet somehow possesses about 60 percent of the angular momentum. But the Sun, which possesses 99 percent of the mass, has only about 1 percent of the angular momentum. In fact, as you look at Jupiter, Saturn, Uranus, and Neptune, these four planets between them hold more than 98 percent of the angular momentum in the entire solar system. This is obviously a result of them having been "birthed" by the Sun in the solar fission process that Dr. van Flandern advocated. As the Sun gave birth to all the planets, like a great cosmic mother, she also gave away most of her own spin energy to them. The biggest ones (the gas giants) got the most, the smaller terrestrial planets like Earth got a bit less, and in the end this process allowed her to calm herself, spin less intensely, fire off fewer deadly solar flares, burn in a calmer and more consistent way, and create the "habitable zone" we talked about earlier—a place where higher forms of life like ours could live and flourish. It truly is a beautiful, harmonic, and balanced view of how things came to be, almost poetic in its perfection and elegance.

So I ask you here: Which view makes more sense? The idea that the Sun gave birth to all that we see and experience, at least here in our little corner of the universe, or that it all came together randomly, as the accretion model suggests? For me, I choose the beauty. I choose the idea that we are all part of very big and very beautiful Solar family.

Okay, back to physics.

Interestingly, these four planets also are unique in another interesting way that relates to the notion that spin energy equates to other types of energy, like heat. Jupiter, Saturn, Neptune, and Uranus are the only planets in the solar system that are somehow radiating more heat energy out then they are receiving in from the Sun. Somehow, they are behaving like mini stars, and putting out energy that cannot be accounted for in any conventional models.[6]

Now, in those same conventional physics models, spin energy doesn't amount to anything, at least in terms of how the planets might interact with each other. But as we learned earlier, it is spin that creates a portal, or a gateway or a star-gate—whatever you want to call it—through which energy enters our universe. To me it is no surprise then that Nelson found that the four planets with the most spin energy had the greatest physical influence on the Sun. But the implications of Nelson's work go waaaay beyond that.

In 2008, the Astronomical Society of Australia published a paper by three astronomers titled "Does a Spin–Orbit Coupling Between the Sun and the Jovian Planets Govern the Solar Cycle?"[7] The paper claimed that they had found a link between the rotation of the equatorial region of the Sun and the Sun's orbital rotation around the barycenter of the solar system (think of it as the center of gravity). They went on to state that "this synchronization is indicative of a spin–orbit coupling mechanism operating between the Jovian planets and the Sun." In plain English, that means that there was some kind of synchronization or symbiotic relationship between the Sun, Jupiter, and Saturn that was actually *driving the sunspot cycles*. Now again, there is absolutely nothing in the conventional view of physics which can account for such a relationship. They even acknowledge as much in in the abstract (summary) of their paper: "*However, we are unable to suggest a plausible underlying physical cause for the coupling...*"[8]

So this goes even beyond what Nelson found, that the planets had simply an influence on physical reality here on Earth. What the Australian paper shows is that conventional physics has it upside down: The Sun doesn't rule the roost because of its size or gravitational and magnetic influences; the planets are the dominant force in our solar system because they possess almost all the (hyper-dimensional) spin energy.

To put it simply, the tail wags the dog, not the other way around. It is spin energy that matters the most, not gravity. Not only that, but the planets and the Sun are somehow linked by an unseen Force that somehow propagates through the emptiness of space without being detected. We already know, of course, how the unseen force propagates: It passes through the higher dimensional Aether that we talked about earlier. This makes it appear as though you can have a cause in one place and an effect in another at the same time. Again, this gives the illusion that the unseen force is traversing the vast three-dimensional distances faster than the speed of light when in fact it isn't. What's instead happening is that the disturbance is caused by something changing here in three dimensions, say Jupiter moving into a 120-degree alignment with the Sun. This causes ripples or waves in the Aether, the higher dimensions, and, just like the ripples on a pond, this wave passes through the higher realms and touches the Sun in its higher state. The Sun then reacts in some way, perhaps by spitting out a moderately sized solar flare, and the cause and effect appear to be simultaneous because the "wave" or ripple did not visibly travel the distance in between. So, although nothing may travel faster than light in 3D, there is no reason it can't do so in the fourth dimension (or higher).

Astronomy Drives Astrology

What was even more interesting about Nelson's studies (and other, later studies) was not just that they showed the planets had an influence on the energetic output of the Sun; it was the specific geometric configurations under which they did so. A 1951 article in *Time* magazine described them like this:

Nelson studied the records of RCA's receiving station at Riverhead, N.Y., looking for some correlation between the magnetic storms and the positions of the planets. He found that most of the storms took place when two or more planets were in what he calls a "configuration": that is with angles of 0°, 90°, or 180° between the lines connecting them with the sun. The more planets involved in a configuration, the more serious the storm is likely to be. During the great magnetic storm of July 1946, for instance, three planets (the earth, Jupiter and Saturn) were in a configuration, and three others (Mercury, Venus, and Mars) were also in a "critical relationship."[9]

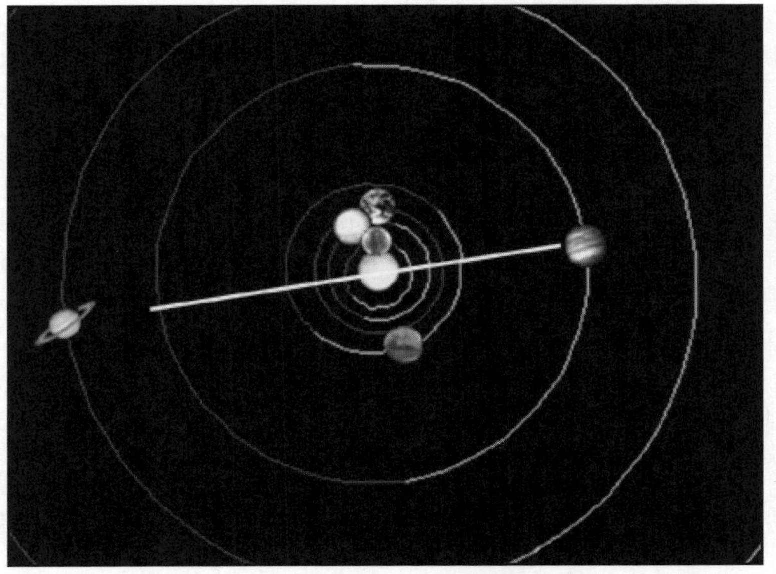

Image 3.2. Jupiter and Saturn in opposition (relative to the Sun), December 21, 2010.

One of the things that immediately caught my eye was the fact that these configurations, as Nelson called them, equated with what are known as "specific aspects" in astrology. For instance, when two planets were either lined up with the Sun but separated by 180 degrees (which is called an "opposition" in astrology) or at a 90-degree angle to each other (which is called a "square"), there would be heavy magnetic disturbances and storms. Conversely, when the planets were at 120 degrees to each other (called a "trine") or at 60 degrees relative to each other (a "sextile"), the Sun was noticeably less active. Using this system, Nelson declared that he could predict the disturbances on the Sun with an accuracy of more than 90 percent.

Now what's interesting is just what these various "aspects" are said to represent in astrology.

An opposition is exactly that. It indicates two or more planets directly opposite each other in the sky (from the viewer's perspective), and it is considered a "tense" aspect, meaning it causes outright conflict between the two planets involved. A square is similar, but not as powerful because the direct opposition has now become a 90-degree alignment. This is also a tense aspect, more akin to friction between the two planets (and what they symbolically represent) rather than outright conflict. The other two configurations, the trine (120 degrees apart on the chart) and the sextile (60 degrees), are harmonious aspects, where everything should flow smoothly and everything should be quiet and peaceful.

So what Nelson found was that these astronomical conditions, two planets at angles of 90 or 180 degrees from each other as an example, exactly correlated to their astrological meanings. Squares and oppositions caused big magnetic storms on the Sun; trines and sextiles made the Sun quiet.[10] I find the implications of this staggering.

What it means is that not only is astrology real—its longtime assumptions can be correlated in the physical—but it is also driven by the astronomy. As the planets move in their orbits about the Sun, the changing geometric configurations and alignments change physical conditions in other locations around the solar system. So not only are the laws of physics not laws at all—they can be changed by the whims (and the orbits) of the planets—but they are a local phenomenon. Just as Tip O'Neill (former speaker of the United States

House of Representatives) once quipped "all politics is local," what Nelson and hyper-dimensional physics show is that *all physics is local,* too. Although the effects may be subtle, the reality is that the laws of physics themselves, things like the passage of time and the speed of light, should be different in every different solar system, because they will all have different mixtures of planets, stars, orbital relationships, and spin energy.

And there is another implication here that we have to address at this point. Remember that the whole notion of astrology is based on the idea that the planets influence not just our physical reality, but our actual consciousness. That they somehow drive our lives in certain directions and place options and opportunities in our path for us to choose our own destiny. What Nelson's study proved was exactly that: The planets do influence our lives, experiences, emotional states, and perhaps even our destinies.

The human brain is nothing but a complex electrical signal transmitter, which is far faster and more complex than any computer. Our brains transmit signals to our bodies, but they also send out (and receive) transmissions in the form of brain waves. The simple truth is that there is really no difference between Nelson's shortwave radio signals and the neurological signals in your brain, which you experience as thoughts and emotional states. Radio signals are a wave, and your thoughts and feelings are waves as well.

But the point is that if radio waves can be affected by these planetary transitions and movements, our *brain waves* can too. After all, they are simply different types of electro-magnetic transmissions, and theoretically equally vulnerable to being influenced by the planets.

Doesn't it also follow then that astrology, so readily dismissed as a pseudoscience by the scientific community, might have some basis in reality? And even more than that, doesn't it even act as a proof that we all have souls, or higher selves? After all, if everything in the universe is generating energy, and every bit of energy is actually coming from higher dimensions, then aren't our thoughts, which are also nothing more than electrical energy, actually coming from higher dimensions, higher planes?

Without realizing it, Nelson had stumbled upon the key to unlocking some of the deepest mysteries of life. His studies could and did form the basis for proving experimentally, by observation, testing, logic, and deduction, that we all are linked to our higher dimensional selves through the Aether. And more than that, he provided a basis for us to understand just why we do the things we do, why at certain times in our lives things seem to flow more easily for us, and why at others they seem to pile up and make our lives more difficult.

Okay, I know all of this sounds really neat, if only you could believe it, right? After all, I'm an engineer, not a physicist, so why should you believe me over brainiacs like Steven Hawking and Michiu Kaku? Well, let me show you.

4 The "Laws" of Physics

Most of us, even if we don't follow science very closely or pay attention to what NASA is up to, have some idea and some curiosity about the nature of the universe we live in. It's hard these days to watch *Oprah* or *The View* or even the nightly news without some aspects of metaphysics creeping into the conversation. I believe that at some level, all of us are curious, at least to a limited extent, even if not all of us are hard-wired to immerse ourselves in the gory details of the latest cosmological debates.

For centuries, really from the time of the earliest civilizations like ancient Egypt and Sumer, questions have been asked about the nature of God's creation. This questioning led to the great philosophers of antiquity, men like Plato and Aristotle, and their musings in turn gave birth to the science of physics. At its heart, despite its mathematical complexity, physics is really nothing more than an effort to understand what everything is made of and how it all works together to create this experience we call physical reality. In recent years, it has become popular to assume that we have it all figured out, and that there is little if any mystery left to unveil. Nothing could be further from the truth.

There are really three basic pillars of modern, mainstream physics: Newton (Newton's laws of motion), Einstein (relativity), and particle, or "quantum" physics, the brainchild of German physicist Max Planck. There is now a popular new idea called string theory, but it's really just hyper-dimensional physics with a mainstream veneer.

When you add up all three of these physical models of the universe, the simple fact is that none of them can explain everything that we see around us. Newton works great as long as nothing rotates. Einstein's relativity works great in two dimensions and at the larger scale, but falls apart at the sub-atomic level. Quantum mechanics works great as long as you don't include gravity in the equations. The simple reality is that all three of these mainstream pillars are fatally flawed.

Newton's Laws of Motion

Sir Issac Newton (1643–1727) was really the first serious physicist in the Western world. His masterwork, *Philosophiæ Naturalis Principia Mathematica,* put forth his so-called "three laws of motion" that sought to explain the rules of gravity and momentum (inertia) in mathematical terms. His three laws, simply stated, were:

1. An object in motion will stay in that same motion unless an outside force acts upon it.
2. A body will accelerate proportionally to the force and inversely proportional to the mass (force equals mass times acceleration).
3. Every action has an equal and opposite reaction.

Newton's laws of motion are certainly one of the cornerstones of modern physics, and often the subject at hand when someone refers to the "laws of physics." But the truth is that Newton's laws only work under certain conditions. Basically, they only work if the object being measured doesn't rotate. That's kind of a large problem, because virtually everything (if not everything period) is constantly in motion, constantly rotating.

For instance, you and everything in your body is rotating as you read these words, wherever you are. You may think you're sitting perfectly still in your easy chair or daybed, but you're not. You're rotating. There are electrons inside your body that are rotating around the nucleus of every atom in it. The Earth itself is rotating on its spin axis. The Earth is rotating (orbiting) around the Sun. The sun is rotating around a giant black hole at the center of the Milky Way galaxy, and the galaxy appears to be rotating around other galaxies in a grouping called a galactic cluster. So it's hard for me to be able to endorse a "law" of physics that doesn't account for rotation.

But beyond that, there are numerous experiments conducted by dozens of rogue researchers through the years that prove that Newton's three laws are *all* wrong. The late Dr. Bruce DePalma (whose brother is well-known Hollywood director Brian DePalma) started his scientific career by graduating from Massachusetts Institute of Technology in 1958. In graduate school, he pursued electrical engineering, both at MIT and Harvard. After grad school, DePalma went to work for some of the nation's most prestigious scientists and scientific institutions.

Early in his career, he became fascinated by the exotic properties of rotating, magnetized gyroscopes. Unfortunately, dabbling in such arcane physics was frowned upon by those in academia, and after almost 20 years of watching the American science establishment up close, DePalma finally got fed up and decided to resign. Free to embrace and extend his growing infatuation with rotation, DePalma carried out the most exhaustive laboratory studies ever conducted on the subject from the 1970s onward until his untimely death in 1997. One of the most straightforward and yet probably the most profound of all his many rotational experiments was called simply "the spinning ball experiment."

Conceptually, the experiment could not have been much cheaper or easier to carry out. Two 1-inch steel balls, like those found in old-school pinball machines, were placed in a test rig. One ball was then spun up to 27,000 RPM. The other ball was in a stationary, non-rotating container. Once the spinning ball reached 27,000 revolutions per minute, DePalma would eject both balls out of the test rig and into the air with a fast, equal, upward motion. This caused both of the pinballs, one rotating and one not, to fly out of their retaining cups in the same upward direction, and with equal upward force ("for every action there is an equal and opposite reaction").

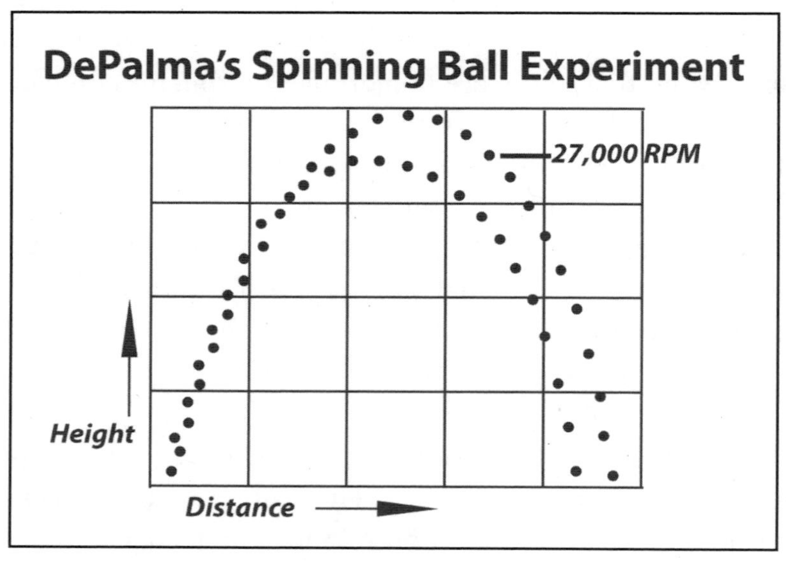

Image 4.1. DePalma's Spinning Ball Experiment. Image by Krys Lilly.

The result shattered Newton's third law of motion. The ball that was spinning rose faster, went farther, and fell at a faster rate of speed than the one that was *not* rotating (so much for "for every action there is an equal and opposite reaction").

Now the simple fact is that what DePalma did in this experiment, which he repeated again and again, is *impossible* according to the accepted laws of physics. Somehow, some way, the act of simply rotating the little steel ball had added energy into it and allowed it to limit the effects of gravity on a small scale. Subsequent experiments DePalma conducted eventually disproved Newton's other two laws as well.

Einstein and Relativity

The next major step forward in physics after Newton was the theory of relativity proposed by Albert Einstein in his paper *On the Electrodynamics of Moving Bodies,* in 1905. In actuality, there are two theories of relativity: special relativity and general relativity.

Einstein's special relativity introduced a number of modifications to classical mechanics (Newton), with mixed results. Special relativity has to do with the structure of what is called "space-time," a mathematical model that combines space and time into a single unified fabric. According to Einstein, space-time is the physical universe we live in, which consists of three spatial dimensions (length, height, and breadth) and one linear dimension of time (that is, time goes in a straight line from past to present to future). The thread that holds space-time together is the idea that the speed of light (186 thousand miles per second) is the maximum speed at which any energy, matter, or information can travel. If it is surpassed it would lead to the destruction of the essential relation between cause and effect. In other words, you can't have an effect before you have a cause, or an effect in less time than light can make the trip, because if you do, the universe falls apart in one big nasty logical paradox. The best way I can put the mainstream view is the way Dr. Carl Sagan once did on *The Tonight Show* with Johnny Carson: "We don't know for a fact that the speed of light can't be exceeded, we just know that a whole lot of things make a whole lot more sense if it isn't."

However, keep in mind that there really isn't any evidence that the speed of light is unsurpassable, and plenty that it is, and the universe has yet to fall apart. Not only are there the experiments done by Cleve Backster that I mentioned previously, but there are also numerous other physics papers that prove the Universe does not have a speed limit. In 2008, physicists at the University of Geneva conducted an experiment in which two "entangled" photons (particles of light) were separated and sent over fiber optic cables to the villages of Satigny and Jussy, which were more than 11 miles apart. Upon arrival, the researchers discovered that the two particles[1] were somehow not only in communication with each other, but that this communication had taken place more than 100,000 times faster than the speed of light. Stunned, the scientists commented that there was just no time for the two photons to have communicated with each other—unless they were somehow connected by a frame of reference "that wasn't readily apparent to the scientists"[2] (hmm, like maybe the "Aether?").

General relativity is a theory of gravitation that Einstein developed later. He created general relativity because classical mechanics fell apart when actual experiments were conducted on objects in "free fall." The problem was that objects in free fall—a state where no force except gravity is acting upon them at all—were observed to accelerate and decelerate relative to other objects. This flatly contradicts Newton's first law: "*An object in motion will tend to stay in that same motion unless an outside force acts upon it.*" If gravity is the only force on an object in free fall, then how can it be attracted by or to another object without the use of magnetism or some other known and academically acceptable force?

In a somewhat desperate attempt to salvage Newton, upon which virtually all of modern physics was based, Einstein proposed that space-time itself must be *curved*. In his view, objects in the vacuum of space "fall" without any measureable force being exerted on them because *space itself* is curved. Kind of like a ball rolling down a hill even though nobody pushed it down the hill, and there is no gravity to *pull* it down the hill, either. The only problem with this notion is that as far as we know, or at least as far as mainstream physicists are willing to admit, such a behavior can only be caused by gravity or magnetic attraction, and both are measureable forces, and they were sure there was no gravity or magnetism acting on the test subjects during the experiments.

See how ridiculous it all is? What Einstein was arguing is that, because objects on a big scale move in a way they are not supposed to, it must be because space itself is "curved." It can't possibly be because something else, some outside force from a higher dimension is acting on them through the Aether, right?

If you are confused right now, I don't blame you. But it gets worse from here on out.

The biggest problem with relativity, besides the fact that there is zero evidence that space itself is actually curved (and there probably never will be), is that it works okay on the macroscopic, or über-large scale, but falls apart on the microscopic, or sub-atomic level.[3] For the behavior of particles smaller than an atom[4] a new theory called "quantum mechanics" or "quantum physics" had to be concocted. Put as simply as I can make it, quantum physics says that nothing can really be observed or believed at the sub-atomic level, because once you observe a little itty-bitty particle like that it changes the nature of the particle simply because you're looking at it.

Huh?

Or something like that. See, no one, not even physicists, really understand, quantum physics, because basically you can make any observation mean anything you want it to mean. In fact, Richard Feynman, the famed CalTech physicist, once said in 1965: "I think I can safely say that nobody understands quantum mechanics." Nice to know you're in good company, isn't it?

Oh, and besides making no sense at all, quantum mechanics only works really well as long as you don't include *gravity* in the mathematical equations.

So let's get this straight: Newton's laws of motion work great as long as you don't rotate stuff, Einstein works great as long as you don't work with anything smaller than an atom, and quantum physics works just fine as long as you pretend there's no such thing as gravity. And still, college professors and academics and even such brilliant luminaries as Bill Nye the science guy will stand up and straight-facedly lecture us all about "the laws of physics."

See why you gave up on science years ago and started searching for your own answers?

Dark Matter and Missing Logic

Not to lay it on too thick here, but in recent years, it's gotten even worse. As cosmologists (you remember, the guys who spend their days thinking about how the universe got started) made more and more measurements of distant galaxies, they started to notice a problem: There wasn't nearly enough gravity to hold everything together. In fact, there was on the order of 70 percent less gravity than was needed to hold the universe together.

See, the problem is that all the solid stuff—rocks, people, planets, stars, and so on—is all made up of something called baryonic matter.[5] Baryonic matter basically is anything made up of atoms; in other words anything solid. Matter has mass, weight, and so forth, and that is necessary to create gravity. Although no one can actually say what gravity is, we have established that there is a relationship between mass and gravity. The more mass something has, the more gravity it seems to generate. So it stands to reason, then, that, in order for really big stuff, like galaxies, to hold together, there has to be x-number of big solid objects, like stars and planets, to generate all that gravity. But as they made their measurements, the astronomers discovered that there wasn't enough matter in the observable universe to account for the gravity that seemed to holding everything together. The only way the physicists could make it all work was to conclude that there was some other unknown outside force holding everything together. The problem was that this "outside force" would have to come from somewhere *outside*, like, you know…the *fourth physical dimension.*

And that idea is someplace their minds simply cannot go.

So what then did all the scientists and physicists and cosmologists and even Bill Nye the science guy do? Did they all get together and decide that this whole "laws of physics" thing was a big joke and they should just start all over? Of course not. They decided instead that all that missing mass must be out there after all—we just can't see it! It's invisible! Because of this, they decided to give it a name: dark matter. And then they spent a ton of your tax dollars setting out to prove that "dark matter" exists.

As you deal with your feelings about the fact that most mainstream physicists are actually blithering idiots, keep in mind one of the basic axioms of life: Education and intelligence are two completely different things. Since the 1960s, when the far left began to dominate our higher academic institutions, science students have been increasingly taught *what* to think rather than *how* to think. Sadly, this academic brainwashing has led otherwise-intelligent people, mostly physicists and astronomers, to take increasingly embarrassing positions in defense of the "laws" of physics that they learned in schools and universities. Because they are stubbornly invested in Newton and Einstein, they simply cannot consider the possibility of other dimensions. If they do so, they face ridicule from their colleagues, the defamation of their Wikipedia bios, and the loss of their research grants for pursuing "pseudoscience." As a result, they have to hide behind terms like the *quantum medium* to invoke the Aether, *quantum time* to account for cause and effect, and *quantum reality* to explain uncomfortable experimental results. Basically, anytime you hear someone use the word *quantum* to describe anything, you can safely assume he/she doesn't really know how to explain whatever it is he's talking about and is using a buzz word to make you think he does.

This complete inability to acknowledge the hyper-dimensional nature of reality (not to mention the hyper-dimensional nature of *Man*) has led the quantum physicists to make some even more embarrassing proclamations over the years....

Fictitious forces

From very early on, it has been taboo to discuss anything in physics that requires what the physicists call a "fictitious force," something akin to the outside force I just mentioned. In physics, a "fictitious force" is anything that acts on an object in what is called a "non-inertial frame of reference." Okay, so what the heck is that? Well, it's actually a bit easier to explain a true "inertial frame of reference" first.

An inertial frame of reference is one in which the frame of reference itself is in constant motion, constant acceleration, or completely at rest. In other words, anyone inside a box—say, a large shipping container—would get no sense of motion if the container moved in one direction and did not change speed after reaching some specific velocity. So under these conditions, Newton's "laws of motion" work great. But what happens if the ship the container is on suddenly has to apply the brakes and turn hard to starboard to avoid an iceberg? In that case, someone inside the shipping container is going to feel himself pushed toward the front of the container, and he's going to lean toward his left as the ship makes a hard right. Once he begins to feel these forces acting on him inside his little container box, he has now entered a "non-inertial frame of reference," because the forces he feels acting on him are considered "fictitious."

I'm not kidding.

The list of fictitious forces is actually pretty frightening when you get right down to it. Among the listing of heretical forces is acceleration (in other words, the "laws of physics" only work if you don't push down on the accelerator pedal of your car), centrifugal force (caused by spinning, like those big "round-up" rides at the county fair), and two rather more exotic forces, the coriolis force and something called the euler force. Oh, and there's one other "fictitious force" we should be on the lookout for if we want to keep the laws of physics preserved for future generations: gravity.

Again, I'm not kidding.

So just to clarify, the "laws of physics,"—as defined by Newton, Einstein, Planck—work just great as long as you don't account for gravity, rotation, centrifugal force, or acceleration. And don't forget: They also work just great if an object is completely at rest, which is defined as either moving with constant velocity or standing completely still (remind me again how something can be "completely at rest" if it is also "moving with a constant velocity" at the same time?) And then there's the part where Einstein noticed that nothing ever stands completely still or moves with constant unchanging motion in one direction. But that was because space-time is curved, right?

Oh, forget it!

The reason scientists must exclude higher dimensions from the science of physics is because it brings the scientists uncomfortably close to confronting that of which they cannot speak: God. Because of the historical conflict between science and religion, astrophysicists are simply incapable of acknowledging higher realms, because that then implies a *maker* of these higher realms. So is it really any wonder that science can't really explain how things got to be as they are? That's always what happens when you remove God from the equation.

In simple terms, if we had followed Maxwell's original vision of God's creation instead of Heaviside's censored sub-set, we would have never fallen into the absurd intellectual mouse trap of "fictitious forces," dark matter, or a fixed speed of light. We wouldn't need to, because we'd *know*—and most likely years ago would have *proven*—that God himself was holding the universe together from these higher realms.

It was with all that in mind, with this theoretical underpinning as my foundation, that I began to look at the astronomy and the astrology of the current day and the coming age. And it was from that perspective that I discovered that somebody else, a long time ago, during a period when we perhaps spent less time in our minds and more in our hearts, had figured all of this out intuitively. And not only that, they wrote it down, inscribed it on their monuments, and left it as gift for us to re-discover today.

We have a chance, right now, during this pivotal period in our history as a species, to reclaim that which is the greatest gift the Universe has given us: our hearts.

5
The Physics of Time

Somehow, it seems that a number of spiritually advanced ancient cultures—the Mayans, Egyptians, and Indo-Aryan Hindus of South Asia at the least—understood the fundamental concepts of higher dimensions and the higher dimensional nature of Man we have just covered. They were among the first civilizations to mark the passage of time with calendars, and they seem to have used them for far more than just calculating the rainy seasons or the annual flooding of the Nile. How they are different from the way we perceive time today is that each of these ancient cultures viewed time as cyclical as opposed to linear. Today, we speak of time as if it is one straight line from the Big Bang to the present day, with each passing moment lost forever in the mists of history, never to be repeated, never to be reclaimed, never to be relived. In the Western traditions, we are taught that each of us only lives once, never to be born again, and that we are under the constant vigilance and influence of a judgmental God. In short, in the modern West, you are one and done, so you'd better get it right.

But in each of the ancient cultures I mentioned, time and life were viewed in a much different way. To them, life, the universe and everything in it were continuously repeating cycles. The Earth, the Sun, and the stars were all in constant motion, always changing but over eons coming around again and again to the same places, the same circumstances, the same astronomical alignments. Even our own souls, viewed as immortal and unceasingly growing, returned again and again in a constant cycle of reincarnation into physical bodies to learn and live and grow.

In short, in the west, time is an arrow. To the ancients, time is a wheel. These contrasting world views are reflected in the specific and mathematical ways that we measure time: by our calendars.

What most of us don't appreciate is that all calendars are inherently astronomical, calculated by the positions of objects in the heavens, be it the Sun, the stars, or the planets. To us, they are just numbers on a sheet of paper or in the lower righthand corner of our computer screens, and they are useful to us only in the sense that they give us a means of recording exactly who said or did something that we deem worth recording, and "when" they said or did it. We rarely consider that these dates and times are based purely on the positions of the planets and the stars at any given moment, and even more rarely whether such alignments might have an impact on our daily lives and even our physical reality.

In our Western calendar, the date is calculated from the location of the Earth in its annual orbit around the Sun. In order to do this, the exact length of a day as well as the exact length of a year must be identified, and that is not such an easy thing to do. For instance, there are two different ways to calculate the length of a day: the sidereal method and the solar method. The sidereal method calculates the day with respect to fixed stars, and the solar method calculates the day with respect to the time it takes for the Sun to return to the same exact position in the sky on consecutive days.

Our modern calendar, called the Gregorian calendar, came into being in 1582 after Pope Gregory the 13th issued a Papal Bull (order) to reform the previous calendar, the Julian, created by Julius Caesar of Rome.[1] The Julian calendar had become increasingly inaccurate, and by the year 1582 many key religious celebrations, like Easter, had drifted far from the dates they were originally celebrated. The Julian calendar had been adopted in the Gregorian year of 46 BC after Caesar had spent considerable time consulting with the Egyptian-educated astronomer and priest Sosigines.[2] Caesar had gone to Egypt in 48 BC in pursuit of his friend and rival Pompeii after defeating him in the battle of Pharsalus and ending a Roman civil war. While there, Caesar was educated in the ways of Egyptian astronomy and religion by the Egyptian Queen, Cleopatra.

What is less clear is Caesar's motive for creating a calendar system in the first place. Although it was inaccurate, the Roman lunar calendar was hardly a logical concern for a warrior like Caesar. He was obsessed at that point in his life with becoming the first emperor of the Roman Empire, and creating a new calendar was hardly the type of activity that would, on the surface anyway, advance such ambitions.

What we tend to forget in our modern perspective on such things is that the men of his age were not merely politicians or soldiers or even both, but frequently priests, mystics, and magicians as well. Most of them had a belief that their paths had been shaped by divine and mysterious forces, and they spent a great deal of time trying to master and understand these forces. Caesar would have been no different, and it is likely that Cleopatra introduced him to the secrets of Egyptian ritual and magic as a payment for his gift of placing her exclusively on the Egyptian throne over her brother Ptolemy VIII. Her ambitions rivaled his, and showing Caesar the powers hidden in the rites and science of the world's oldest civilization would have been a tremendous bonding agent between them.

So why then did Caesar institute a calendar based on what would become the traditional Western, linear view of time? The Egyptian system, based on the annual inundation of the Nile valley in the month of July, was cyclical. Perhaps the reason was that Caesar understood, as a conqueror, that men who knew they would live over and over again (an inherent implication of cyclical time) would have less fear of death in battle, and hence would not fight as hard to stay alive and thus be useful to Caesar and to Rome.

In any event, the ancient Egyptian stellar religion is, like the Mayan and Hindu, based on the notion of cyclical time, and it is centered on the movements of certain specific stars in the night sky. According to authors Graham Hancock and Robert Bauval, the Egyptians were obsessed with the movements of two specific constellations: Orion, with its famous "belt stars" *Alnitak*, *Alnilam,* and *Mintaka*, and with Canis Major, which contains Sirius, the brightest star visible in the night sky. They were so fascinated with these two collections of stars that they deified them as the goddess Isis (Sirius) and god Osiris (Orion), and they built permanent monuments to them in the form of the great pyramids of Giza, which reflect the layout of the three stars of Orion's belt.

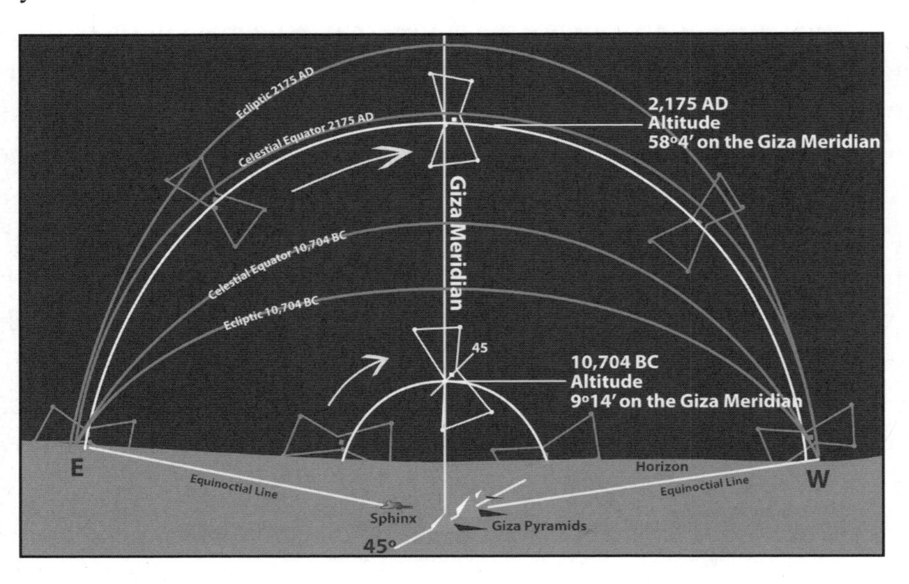

Image 5.1. View of the meridian from the Great pyramid of Giza on the winter solstice. As the years pass, Orion's belt stars seem to rise and fall in 13,000 year cycles due to the effects of precession. Image by Krys Lilly.

Hancock, in his book *Heaven's Mirror*, made a convincing case that the pyramids in general and the Great Pyramid specifically were created as giant "meridian machines," devices used to monitor the movements of the heavens relative to the Giza meridian.[3] More specifically, Hancock argued that the Great Pyramid was set up to monitor the movements of the belt stars of Orion as they moved up and down in the night sky, like passengers on some celestial elevator. Because the stars, like the Earth and everything else in the universe, are always moving, this subtle up and down motion must be monitored from the same location and on the same day every year to get an accurate reading. As the belt stars cross the meridian on a specific date each year—say, the winter solstice (December 21st)—the Egyptian observers would have seen it get gradually higher every year on that date. After 13,000 years or so, Orion's belt would then start to descend back down the celestial elevator, until it reached its nadir, or the bottom floor, as it were.

This effect is caused by a phenomenon called *precession*. Because the Earth wobbles on its spin axis like a top that has lost some of its momentum, the "tip" of the axis, the location in space where it theoretically points, cuts an imaginary great circle in the sky. It takes about 72 years for the axis to move about 1 degree along this great 360-degree circle, and at that pace it takes roughly 26,000 years for the Earth's spin axis to make one complete circle in the sky. This 26,00-year cycle of precession is sometimes called "the great astronomical year."

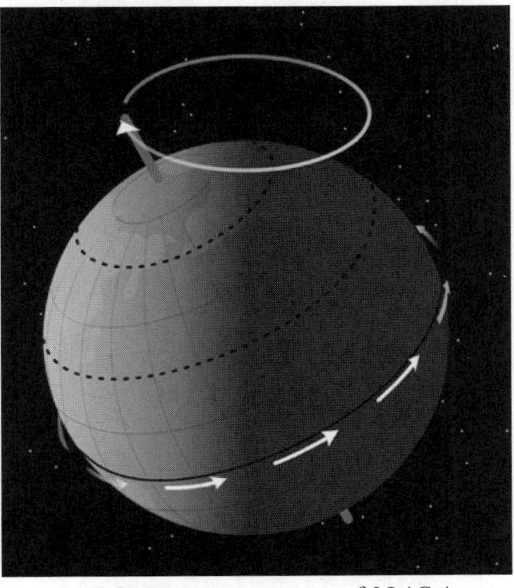

Image 5.2. Image courtesy of NASA.

One of the effects of this phenomenon is that it makes the constellations in the night sky rotate slowly in a counter-clockwise direction when viewed from Earth. This is commonly called "the precession of the equinoxes," and the 12 constellations that are on the plane of the ecliptic (the Sun's equator) are referred to as the signs of the zodiac, and make up the basis for the astrological signs. Because the Earth's spin axis is angled at 23.5 degrees to the Sun's orbital plane, these 12 "houses" of the zodiac (Aries, Taurus, Gemini, Cancer, Leo, Virgo, Libra, Scorpio, Sagittarius, Capricorn, Aquarius, and Pisces) appear to rotate backward across the night sky at that same angle.

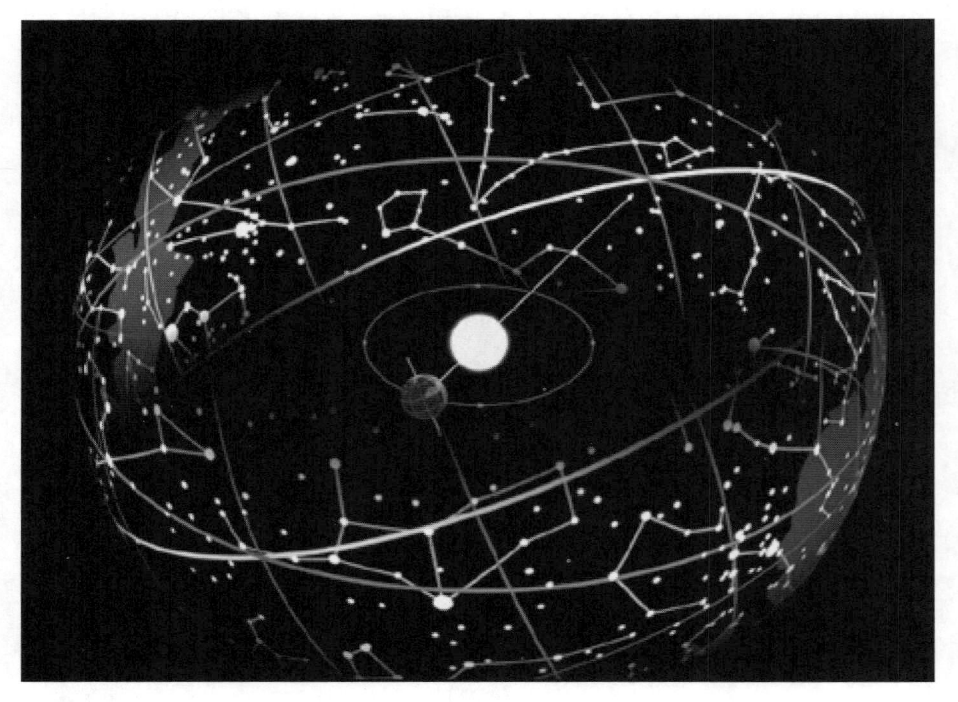

Image 5.3. The precession of the equinoxes. The constellations of the astrological zodiac span the Sun's orbital plane and are therefore at an angle of 23.5 degrees from an earthbound observer's viewpoint. Arrow points to the vernal equinox point, where the plane of the ecliptic and the celestial equator cross in Pisces/Aquarius. Image by Krys Lilly.

These 12 constellations are then measured against the position of the vernal equinox point, the point in the sky where the plane of the ecliptic (the Sun's equatorial plane, and the celestial equator (the Earth's equatorial plane) cross on the spring equinox (March 21st) each year. As the decades progress, this position is seen to be moving slowly backward, and approximately once every 2,160 years, the vernal equinox point moves from one zodiacal "house" to another. Presently, it is generally agreed that we are living in the astrological age of Pisces, but we are also considered to be on the brink of the new age of Aquarius. At 2,160 years each, these 12 houses of the zodiac add up to a precessional "year" of 25,920 years, commonly rounded up to 26,000 years.

Precession, by the way, appears also to have a hyper-dimensional cause. The mainstream thinking is that the Moon's gravity tugs on the Earth's bulging equator (which it demonstrably does) and that this torque force somehow pulls the Earth just slightly off kilter as it spins. However, there are a couple of logical problems with the idea that precession is caused by simple gravitational effects. First, the Moon also precesses, but far less dramatically than the Earth does. Given that the Earth has a far stronger gravitational field, it seems illogical that it would have less of a precessional influence on the Moon than the Moon does on the Earth. Beyond that, there is also the not-so-minor problem of the fact that the Moon is slowly receding from the Earth. Laser reflectors left by the Apollo astronauts on the surface of the Moon (yes, Virginia, we really went—see *Dark Mission*) have shown conclusively that the Moon is moving away from the Earth at the rate of about 1.5 inches per year. In that case, if gravity is the cause of precession, then the Earth's rate of precession should be slowing down. Instead, as the Moon recedes, the rate of precession is actually speeding up. It's now thought to be 25,828 years, as opposed to the official 25,920 (360 degrees times the 72 years it takes to move one degree along the precessional arc).

But I digress. The point is that Hancock and his coauthor, Robert Bauval, argued in their book *The Message of the Sphinx* that these pyramidal Meridian machines were used to track the progress of Orion's belt stars as they rose and fell from the perspective of the Giza plateau in the endless cycle of precession. For approximately 12,882 years, Orion could be seen to be slowly descending as viewed from Giza, finally reaching its "nadir," the lowest point in the half-precessional cycle, in the year 10,704 BC. From that low point, 9 degrees 14 minutes above the horizon, it has been steadily ascending for the last 12,714 years and will finally reach its highest point, and the completion of the current precessional cycle, in the year 2175. So the idea is that the Egyptians evidently had a long count cycle of time, linked to the phenomenon of precession, which they marked by the rising and falling of Orion's belt over the Giza plateau. Hancock himself declared in the documentary film version of Heaven's Mirror: "I have concluded that the entire Giza plateau is nothing less than a giant precessional clock, counting down to some kind of event. But I have no idea what."

There is little question that the ancient Egyptians understood the mathematics of precession, for embedded in the myths surrounding their most sacred gods, Isis and Osiris (Sirius and Orion), there are numerous mathematical clues to it. At Abydos in Egypt, a great and ancient temple to Osiris still stands today. The Osirion, as it is named, is usually said to have been built during the reign of Seti I (1290–1279 BC), but there are numerous clues that it is far older than that. The architecture bears no resemblance to later architecture in the surrounding temples (it does not even feature any hieroglyphs) and it is actually below the water level for the later temples, meaning it must have been built before Seti's reign. Inside the section that Seti built is the so-called Corridor of Kings, which lists every ruler of ancient Egypt: Gods, demi-gods, and Men, going back more than 10,000 years. Osiris, according to many ancient texts, was the first of the God-rulers.

According to the myths, Osiris was a much beloved and generous ruler of Egypt, but he was murdered by his jealous brother Set. Resurrected by his wife, the sorceress Isis, Osirs went off to rule in the *Duat*, the portion of the sky between Sirius and Orion, basically directly opposite the center of the Milky Way galaxy in Sagittarius. From there, he will rule as judge of the dead by weighing each dead man's heart on a scale against a feather. Someday, the legends say, he will be called to return to Earth and initiate a new golden age that will rebirth the *Zep Tepi*, the "first time," when he ruled Egypt in peace and prosperity.

It is now fairly commonly accepted that these myths form not only the basis of the rituals of Freemasonry, but they are loaded with mathematical and symbolic references to precession. The Djed Pillar, which was taken from a tree raised overnight by Osiris's body when it arrived in the kingdom of Byblos, is said to be the axial pillar that holds up the world and keeps the heavens spinning in a stable manner through the night sky. The number of conspirators against Osiris is 72 (again, it takes 72 years for the earth's spin axis to move 1 degree along the great circle of precession), and the numbers 2,160 and 25,920 appear in numerous ancient texts as well. Jane B. Sellers, an Egyptologist with more than 60 years of study of both the Osiris myths and the Pyramid Texts themselves stated:

> "I am convinced that for ancient man, the numbers 72...2160 ...25,920 [appearing in these texts]...all signified the concept of the Eternal Return."[4]

By the "Eternal Return" she meant the time when the Great Astronomical Year of the Precession comes full cycle, that is, Orion's belt "returns" to its celestial starting point. So the Egyptians seem to be tracking precession so they will know the time when Osiris will return, judge each man by his works, and then establish a new golden age under his rule.

The ancient Egyptians, however, were not the only ones who considered the cyclical nature of time and the precession of the equinoxes to be important, and for very similar reasons. Across the world, there were many equally "primitive" cultures that were tracking it as well.

6 The Golden Alignment

Like the Egyptians before them, the Hindu cultures of south Asia saw time as a great circle. The Vedic religious culture of northern India was the root of modern-day Hinduism, and their beliefs, written down in the form of Sanskrit scriptures called the Vedas, form much of the basis of that modern religion. Within the Vedas are laid out the concept of the Yugas, or ages, which govern the evolution of the world and the consciousness of Man. According to the Vedas, mankind and the Earth pass through four continuously repeating cycles of consciousness that are dictated to a very great extent by the motions of the Earth around the Sun and the motions of the solar system around some undefined much larger central sun, probably the galactic center. This central sun is sometimes called the *Vishnu Nabhi*, and is the source of universal intelligence and spirit. During the Sun's journey around this source, the influence of Vishnu Nabhi waxes and wanes, much like the four seasons of the Earth. When the Sun comes into direct alignment with the Vishnu Nabhi, *Dharma,* the practice of religion and spiritual enlightenment, becomes the overwhelming aspect of human existence. When it is farthest away from Vishnu Nabhi, Dharma fades and the acquisition of material wealth and dominance over others becomes the all-encompassing nature of most men. Dharma is often symbolically represented in the scriptures as a bull. According to the traditional method of calculating the Yuga Cycles, it takes approximately 4 million, 320 thousand years to complete one orbit around the Vishnu Nabhi, and thus one complete cycle of consciousness.

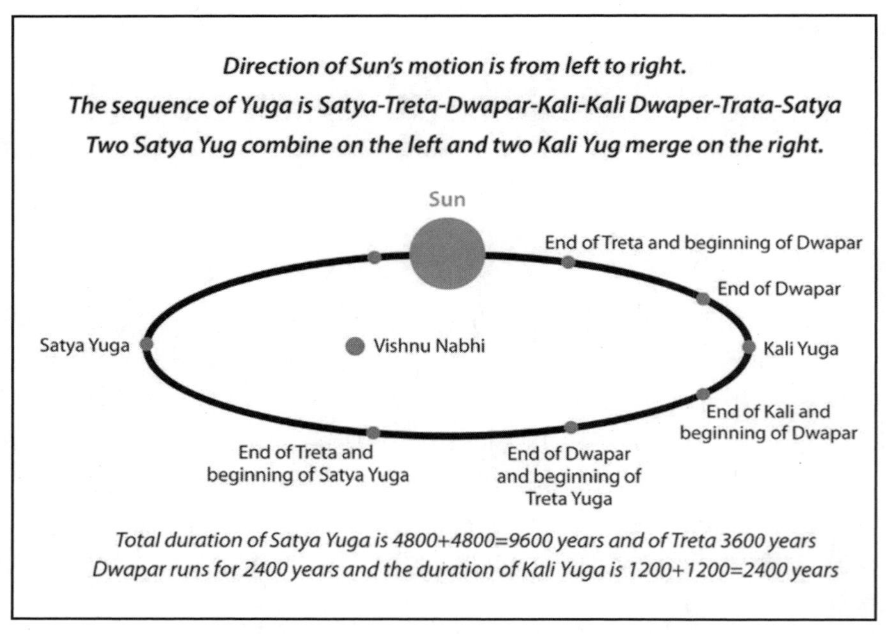

Image 6.1. The solar system's motion around the Vishnu Nabhi and its subsequent effects on human consciousness. Image by Krys Lilly.

There are four phases of the Yuga, and these four Yugas, once completed, are said to equate to one day in the lifetime of an entity named Brahma, who is one of three deities that share the characteristics of the single Judeo-Christian God. In the Hindu traditions, Brahma is the creator, Vishnu the preserver, and Shiva the destroyer (this is very similar to the G.O.D. concept: Generator, Operator, Destroyer).

The Yugas are generally equivalent to Greek concepts of the four ages, the Golden, Silver, Bronze, and Iron. In the Greek, the Earth is presided over by a deity named Astraea during the Gold and Silver ages, but she flees the wickedness of Man in the Bronze and Iron ages. The primary difference between the Greek Ages and the Yugas is the Vedic concept of circular time, as opposed to the Greek view of time as linear. In other words, in the Greek, Astraea is gone, and she ain't coming back. In the Vedic, the golden age of peace is coming once again.

The Vedas define the four Yuga Ages as:

Satya Yuga—The Golden Age.

Treta Yuga—The Sliver Age.

Dwarpa Yuga—The Bronze Age.

Kali Yuga—The Iron Age.

Image 6.2. The Yuga Cycles. The Golden Age is split into ascending and descending aspects, as is the Iron Age. The Silver and Bronze ages each take place separately, on the ascending and descending aspects of the cycles. Image courtesy of Walter Crittenden, after Sri Yukteswar. Used with permission.

Because of the fact that the Yugas are a circle, the two opposite aspects, the Satya and Kali Yugas, at the top and bottom of the "clock face" as it were, both straddle the ascending and descending periods of the cycles. In other words, half of the Golden Age is spent in the ascending aspects of the cycle, and half in the descending. Likewise for the Iron Age, or Kali Yuga. Half of this Dark Age is spent descending to the depths of Kali Yuga, and half is spent coming out of it. If we start the clock at midnight, for instance, then we begin the cycle in the midst of a Golden Age, or Satya Yuga. We then pass through the Silver Age, eventually descending to the Bronze and Iron ages. Half way through the Iron Age, we hit 6 o'clock and begin to ascend again, eventually reaching the top of the cycle once again at 12 midnight. At that moment, Brahma begins another day and the cycle starts up again.

According to various sources,[1] each Age is characterized by specific trends in human behavior physical longevity, and lasts for a specified length of time:

Satya-Yuga: 1,728,000 human years. The Golden Age. Brahma the bull stands on all four legs. The longest of the Yugas, this is the ideal age, characterized by virtue, wisdom, religion, and practically no vice or ignorance. Humans do not hate or envy each other, nor do they ever feel anxious, fearful, or threatened. They solely worship the one true God, hear the one Veda, obey the one law, and practice the one religious process: meditation on the spirit. There is no want, because humans are in a state of full co-creation with God. People live for about 100,000 years.

Treta-Yuga: 1,296,000 human years. The Silver Age. Brahma stands on three legs. Vice is introduced. The good qualities that humans embraced in the Satya-Yuga decline by one third. People introduce religious rites, sacrifices, and ceremonies. They start to act with selfish desires, expecting a reward for their work and religious activities. They live for a maximum of 10,000 years.

Dwapara-Yuga: 864,000 human years. The Bronze Age. Brahma stands on two legs. Spiritual integrity and selflessness are only half of what they are in Satya-Yuga. The Vedas (scriptures) are divided into four parts, and only a few people study them. Sensual desires and diseases begin to well up, and injustice spreads in human civilization. People live for a maximum of 1,000 years.

Kali Yuga: 432,000 human years. The Iron Age. The most degraded of the four ages (Kali literally means "quarrel and hypocrisy"). Brahma the bull stands on only one leg. Only one fourth of human spirit remains and gradually reduces to nothing as the Age progresses. In this Age men are short lived and have less intelligence. They are lax in performing their spiritual duties and exceedingly slow to surrender to God. They are misled, frustrated, and, above all, never truly at peace. The qualities of religion (truthfulness, cleanliness, forbearance, and mercy) and the qualities of life (intelligence, duration of life, and bodily strength and beauty) all diminish. The maximum duration of human life is 100 years, and even that is rare.

If we total all these up, one of Brahma's days, one complete Yuga cycle, lasts for 4 million, 320 thousand years. That's quite a long time, especially if you are stuck in the Kali Yuga without any hope of you or anyone else ever getting out of it your lifetime, even if you have more than one! Because it is traditionally accepted that the current Kali Yuga officially began on February 18, 3102 BC, it would seem we will be mired in it for quite a long time. However, the length of the cycles and one day for Brahma has become the subject of much debate through the years.

An alternative view of the Yuga cycle time scale was put forth by the late-19th- and early-20th-century Indian yogi and astronomer Swami Sri Yukteswar Giri. Sri Yukteswar was, among other things, the guru of Paramahansa Yogananda, who founded the Self Realization Fellowship in California in the 1940s. In his 1894 book, *The Holy Science*, Sri Yukteswar calculated that the descending phase of the Satya (Golden) Yuga actually lasts 4,800 years, the Treta (Silver) Yuga 3,600 years, Dwapara (Bronze) Yuga 2,400 years, and the Kali (Iron) Yuga 1,200 years. The ascending phase of the Kali Yuga then begins, also lasting 1,200 years. Adding all the ascending and descending ages up, he comes up with a grand total of 24,000 years for a single Yuga cycle to complete. Still a long time, but far less than the 4-million-year (plus) total that has traditionally been believed. Sri Yukteswar based his calculations on historical records that indicated that an Indian ruler named Maharaja Yudhisthira realized that the worst times of the dark ages were coming, and he fled to the Himalayan Mountains along with all of his wise men. Resultantly, his court was bereft of anyone capable of accurately interpreting the scriptures that had been left behind, and incorrect calculations resulting in a Yuga scale of millions of years instead of thousands thus took hold.

Yukteswar further argued that just as the cycle of day and night and the passing of the seasons are caused by the celestial motion of the Earth spinning on its axis, so too were the "seasons" of the Yuga cycles caused by a celestial motion; in this case, the motion of the entire solar system relative to the position of the Vishnu Nabhi. His total number of years for the Yuga cycles came to 24,000 years, very close to the 25,920 year (canonical) estimation of the precession of the equinoxes. He was the first to claim that precession was caused not by the gravity of the Moon (which as we've seen it cannot be), but rather by the tug of this distant star. Because that path is elliptical rather than circular, sometimes we are closer to the source (Vishnu Nabhi) and sometimes we are farther away. When we are closer, he explained, life is better:

> The sun also has another motion by which it revolves round a grand center called Bishnunavi which is the seat of the creative power Brahma, the universal magnetism. Brahma regulates Dharma, the mental virtues of the internal world. When the sun in its revolution round its dual come to the place nearest to this grand center the seat of Brahma (an event which takes place when the autumnal equinox comes to the first point of Aries) Dharma the mental virtue becomes so much developed that man can easily comprehend all, even the mysteries of Spirit.[2]

Based on his interpretation of the passing of the Yuga cycles, Yukteswar argued that man was not in the middle of the Kali Yuga, or Darkest Age, but rather in the ascending node of the Dwarpa Yuga, the Bronze Age:

> The position of the world in the Dwapara Sandhi era at present (AD 1894) is not correctly shown in the Hindu Almanacs. The astronomers and astrologers who calculate the almanacs have been guided by wrong annotations of certain Sanskrit scholars (such as Kullu Bhatta) of the dark age of Kali Yuga, and now maintain that the length of Kali Yuga is 432,000 years, of which 4994 have (in AD 1894) passed away, leaving 427,006 years still remaining. A dark prospect! And fortunately one not true.[3]

Using his own system for the calculation of the Yuga cycles, Yukteswar concluded that the beginning of the last Kali Yuga cycle was in the year 701 BC, as opposed to the traditional dating which puts the start of the current Kali Yuga at February 18, 3102 BC. This period is also considered by many Hindus to be the day that the demi-god Krishna died after being mortally wounded by an arrow (time's arrow). Symbolically, this marked the moment when Man began to believe the lie that time is an arrow (linear) rather than a wheel, or circle. If Yukteswar is correct, then we would have moved into the ascending phase of the Kali Yuga in September of 499. Given his identification of the length of this phase as 1,200 years, that would have placed the beginning of the ascending phase of the Dwapara Yuga, the Bronze Age, as September of 1699. However, I have to disagree with Yukteswar for a number of reasons. From his vantage point in 1894, just coming out of a period of intense interest in all things metaphysical throughout Europe and the United States, it must have seemed to him that things were getting better. But he had yet to see the 20th century....

I think we are still in the ascending portion of the dark Kali Yuga, and the transition we are about to go through will be the true beginning of the next age. For one thing, if we were truly in the Bronze Age, our life-spans, at least according to the ancient texts, should already be exceeding 100 years. In reality, living longer than 100 years, despite all our medical advancements, is still "extremely rare," just as the descriptions of the Kali Yuga dictate. And we are certainly in a period where Man is "especially lazy in performing their spiritual duties and exceedingly slow to surrender to God...."

Let's also take a look at some further portrayals of the characteristics of the Kali Yuga from the Sanskrit epic the *Mahabharata*, dated from 400 BC. It describes the politics of the period like this:

> Rulers will become unreasonable: they will levy taxes unfairly. Rulers will no longer see it as their duty to promote spirituality, or to protect their subjects: they will become a danger to the world. People will start migrating, seeking countries where wheat and barley form the staple food source.

It is very hard for me to argue that this is an accurate description of the times in which we live. Rulers—as an example the current U.S. government—have certainly done all they can to remove the concept of God from our schools and even our language. Anyone who expresses a fervent belief in anything beyond the clay-like existence of the material world is ridiculed and dismissed as a kook by the likes of Michael Moore, Bill Maher, and Keith Olbermann. Any mention of God is essentially forbidden in the courts anymore, and the tax burden in the United States and Europe has pushed numerous governments to the brink of bankruptcy. As to our leaders becoming a danger to the world, you don't have to look much beyond the incompetence of the Clinton administration, which led to 9/11, followed by the equally incompetent Bush administration plunging the world into an ongoing and disastrous war in the Middle East. And as to the migration of peoples, how about the literal invasion of illegal immigrants we face from Mexico at the present moment?

But what the *Mahabharata* has to say about human relationships in the Kali Yuga is even more prophetic:

> Avarice and wrath will be common. Humans will openly display animosity towards each other. Ignorance of dharma will occur. People will have thoughts of murder for no justification and they will see nothing wrong with that mind-set. Lust will be viewed as socially acceptable, and sexual intercourse will be seen as the central requirement of life, with the result that even 13- to 16-year old girls will get pregnant. Sin will increase exponentially, whilst virtue will fade and cease to flourish. People will take vows only to break them soon after. People will become addicted to intoxicating drinks and drugs. Men will find their jobs stressful and will go to retreats to escape their work. Gurus will no longer be respected and their students will attempt to injure them. Their teachings will be insulted and followers of Kama (lust) will wrest control of the mind from all human beings. Brahmins (teachers and teachings of the spirit) will not be learned and honoured, Kshatriyas (defenders of the people and the faith) will not be brave, Vaishyas (merchants) will not be just in dealings and shudras (laborers) will not be honest and humble to their duties and to the other castes.

Quite frankly, I can't think of a better description of the United States of America at the dawn of the 21st century than what I just cited from the vantage point of more than 2,400 years in the past. If this isn't a Dark Age, I don't know what is. Fortunately, what I can also see is that while we are still muddled in the throes of the Kali Yuga, we really aren't that far from emerging from it either. Few could argue that life in general is better for more humans than it has been for the last several thousand years. We are clearly on the upswing toward the Dwapara Yuga. And what I can also see is that there is a science and a spirit that is driving us to this new beginning.

7 The Science of the Cycles

One of the first things I noticed in my research about the Yuga cycles involved the debate regarding their length. As we just learned, whereas the standard interpretations had the cycles taking some 4 million years to complete, Sri Yukteswar had a complete cycle at a much more reasonable 24,000 years. Somehow, I felt that both numbers were linked to precession, which I had come to believe was the key to unlocking the mystery of the current age. But even though Yukteswar's numbers were close (24,000 years is about 92 percent of 25,920 years, the traditionally accepted length of a precessional cycle), they weren't exact, and because he was an astronomer, it was hard for me to justify a solid link to the precessionary cycles. Yukteswar also was adamant that he considered the 24,000-year Vishnu Nabhi cycle to be a separate cycle from that of the precession of the equinoxes. But still, I was unconvinced and decided to look at the traditional 4-million-year cycle a bit more closely.

If you take the traditional interpretation of the cycles in "demigod" years, they add up to some pretty interesting totals. Taking the total of 4,320,000 years of a single cycle and multiplying it by 4 (after all, there are four Yugas in the first place), you get a total of 17,280,000 years. No big deal. But then I had flash. Some cultures (like the Mayans) divide their long count calendar cycles into five parts, not four, so I added another 4,320,000 years to the total. That gave me 21,600,000 years, which I found to be a very intriguing number, not the least of which because it ties back to the length of an Age of the precession of the equinoxes, which are divided into 12 segments of 2,160 years; also 2,160 miles is, as we learned in Chapter 3, the exact diameter of the Moon at its equator. I then decided, just for fun really, to add one more full-length Yuga cycle to the equation. So let's see, 21,600,000+4,320,000 equals... 25 million 920 thousand years. That's right, 25,920,000. Why does this number seem familiar? Because as we learned earlier, one complete cycle of the precession of the equinoxes takes just slightly less than 26 thousand years. Or, to be more exact, 25,920 years.

It was at that moment that I realized that *both* the traditional interpretation of the Yuga cycles *and* Yukteswar's numbers were correct. 25,920 years is a "decimal harmonic" of 25,920,000 years, in the same way that the 21,600,000 number is a decimal harmonic of 2,160. All you have to do is slide the zeros a few places. Now, you can take issue with my decision to add a fifth and sixth sub-cycle to the total of 17, 280,000 years, but the truth is that was not a reach at all. There really are only four states of consciousness in the Yugas—the Golden, Silver, Bronze, and Iron—but there are in fact *six phases* of the Yuga cycles themselves. Take a look again at the Yuga cycle graphic shown first in Chapter 6 (Image 7.1).

In order to make one complete cycle of the Yugas, mankind must pass through *six specific states of consciousness*: one Gold, two Silver, two Bronze, and one Iron. That makes six distinct phases of each 4-million-year-plus Yuga cycle, so to me it is totally defensible to add up six sub-cycles into an even bigger long count cycle.

What this unlikely "numeric coincidence" said to me is that, just as I said at the beginning of this book, everything in God's creation has some kind of design. Yukteswar's numbers, though a little off, were somehow connected to the cycle of precession, and the traditional count numbers linked to some larger cycle that has yet to be figured out. One was a long count calendar of roughly 26 *thousand* years, and one an ultra-Long Count Cycle of roughly 26 *million* years. For the record, Yukteswar stated in *The Holy Science* that the 24,000-year count he came up with was not linked to the precession of the equinoxes, but was an entirely separate cycle of the Sun's orbit around the Vishnu Nabhi. So taking him at his word, there would actually be three cycles embedded in the Yuga's: the 24,000-year Vishnu Nabhi cycle, the 25,920-year cycle of precession, and the 25,920,000-year cycle of the traditional reading of the Yugas. Now the task was to figure out the meaning of these numbers, and how they fit into the grand physical scheme of the Universe.

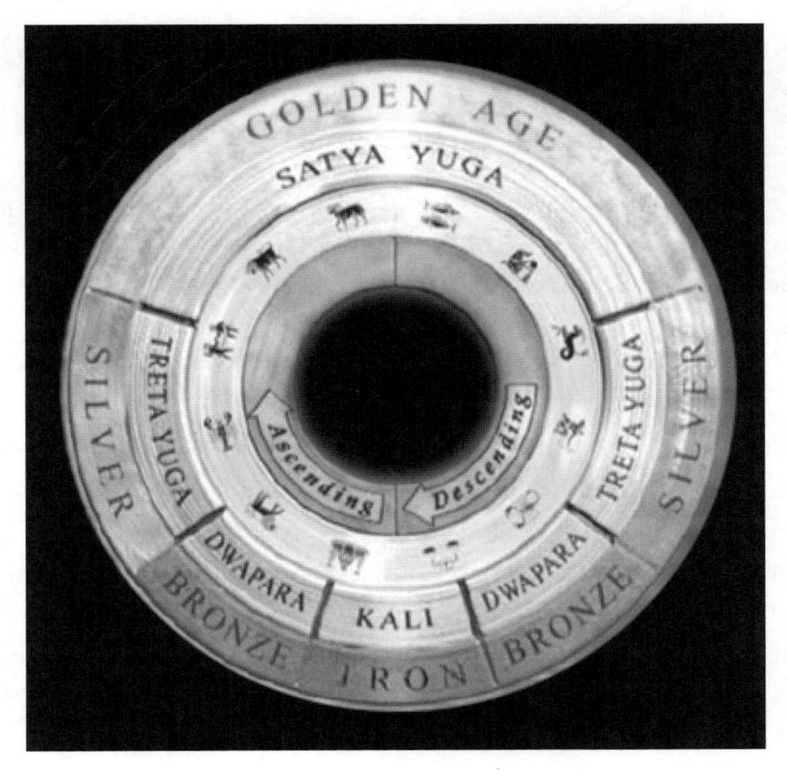

[Image 7.1]

It's just as *The Holy Science* claims: The Earth and with it the entire solar system are, in fact, in orbit around another celestial body, and nobody in mainstream science disputes it. That body is the central core of the Milky Way galaxy of which we are a part. It takes somewhere between 250 and 260 million years for our solar system to complete one orbit around the galactic core, a number that obviously correlates as another possible decimal harmonic of 26. In fact, I'll go on record as saying that, if a more accurate reading of the length of one "galactic year" is ever calculated, it will turn out to be 259,200,000 years. Once again, there is that 25,920 correlation.

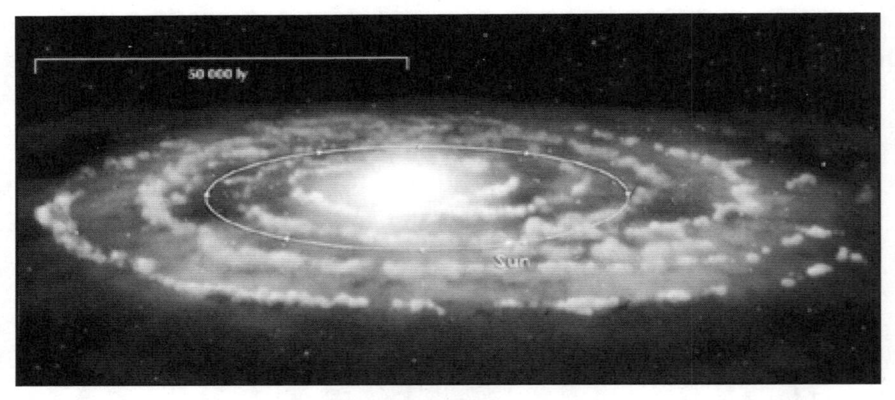

Image 7.2. The solar system's elliptical path around the center of our galaxy. The Sun resides 26,000 light years from the center and one orbit takes between 250 and 260 million years to complete one orbit. Image by Krys Lilly.

But of course, it gets better. Not only does the solar system's orbital period link up numerically to our own local physics and precession, so does the relative distance of our solar system to the galactic core. Care to guess? How about 26 thousand light years from the center of our galaxy?[1] Again, I will lay down a bet right now that if a more accurate determination of our exact position to the center of the galaxy is ever determined, it will in fact turn out to be 25,920 light years, rather than the rounded up 26,000 light years.

So once again, we are confronted with that "26" number. Even more interestingly, out of that 250-million- to 260-million-year cycle, yet another link to the number 26 emerges. In the mid-1980s, two geologists, named Donald Raup and J.J. Sepkoski, Jr., wrote a scientific paper titled *Periodicity of extinctions in the geologic past.* In it, they looked at the geologic fossil evidence of extinctions taking place during a 250-million-year period (obviously very close to one galactic year). What they found was evidence that during that period, large scale extinctions took place on our planet on a regular, repeating basis, and that the interval between extinction events was—you guessed it—roughly 26 million years.[2] Fortunately, the last such event appears to have taken place about 11 million years ago, so we don't really seem to have much to fear regarding this ultra-long count cycle. But it is clear to me that the 26 million year long count Yuga cycle is connected to these periodic extinctions.

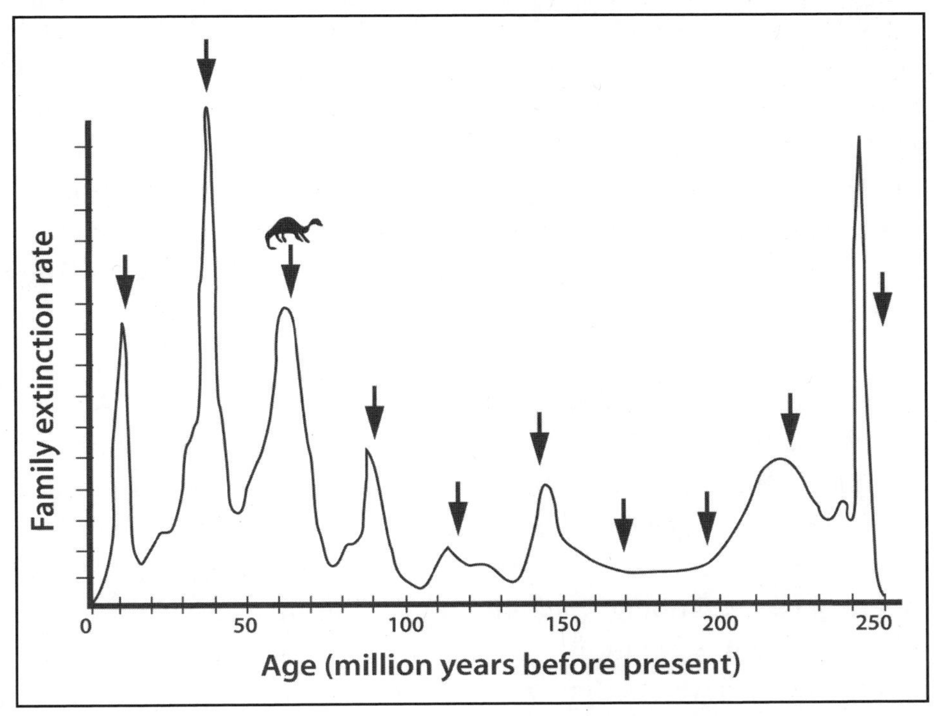

Image 7.3. Chart showing regular intervals of extinction events in the recent geologic past. The dinosaur icon marks the K/T boundary event thought to be linked to the extinction of the dinosaurs 65 million years ago. Image by Krys Lilly.

Had both Sri Yukteswar and the traditional calculators of the length of the Yuga cycles intuitively tapped into some kind of mathematically coded secret that was linked to the precession of the equinoxes and the orbit of the solar system around the galactic center? And if they had, what was that code trying to tell us? Once again, the answer was right in front me.

The core thesis of Yukteswar's book was that our solar system was orbiting some other body that would alternatively raise and lower Man's consciousness based on our proximity to it. He saw us on 24,000-year (I say 25,920-year, but they are close enough) elliptical path around this other body, which he of course named the Vishnu Nabhi. Now, there may well turn out to be such a local companion, and it may well turn out to have an influence on the physics as well as the consciousness of our solar system. But the fact is that the center of the Milky Way fits all of the descriptions of the Vishnu Nabhi to a T. Not only does the solar system orbit the galactic center, but it orbits on an elliptical path that alternately takes us closer and farther away from the core over that 260-million-year orbit. But even more impressively for Yukteswar's thesis, our proximity to the galactic core also does in fact have a measureable effect on our consciousness.

In 1997, Dr. James Spottiswoode of the Cognitive Sciences Laboratory in Palo Alto, California, published a paper entitled "Apparent association between effect size in free response anomalous cognition experiments and local sidereal time." For those of us not versed in the language of academia, "anomalous cognition" means ESP, or Extra Sensory Perception. For decades, Spottiswoode and others have been doing research into ESP using a variety of methods including precognition, card reading, and so forth. Through the years, they've developed a rigorous scientific protocol for measuring "anomalous cognition" and have done a pretty good job documenting it as a statistically valid phenomenon. What Spottiswoode's paper sought to do was to find out if there was any pattern to the more than 20 years of data they had collected. To no one's surprise, there was. To virtually everyone's surprise, it turned out to be linked to the center of the Milky Way galaxy.

What Spottiswoode found was that there was a link between Local Sidereal Time and ESP. The study showed that at high noon LST, anomalous cognition, or ESP, suddenly jumped in intensity. Showing an effect that was way beyond the margin of error, Spottiswoode's study also concluded that the effect was most dramatic at 13:30 (1:30 p.m.) Local Sidereal Time. In other words, ESP got dramatically more pronounced—intuition spiked—between high noon LST and 13:30 LST, peaking at 13:30, before dropping off again to "normal" levels. And what is most significant about that is that 13:30 LST is when the constellation of Sagittarius is most directly overhead at the latitudes where the stuides were conducted. And Sagittarius just also happens to be the location of the center of the Milky Way galaxy, specificially near a star known as Sagittarius A.

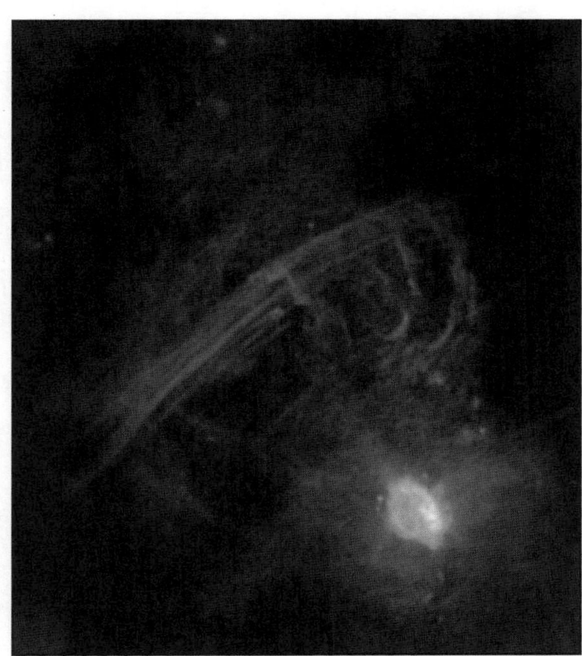

Image 7.4. Chandra X-ray image of Sagittarius A, where a supermassive black hole is thought to lurk. Sagittarius A is ejecting energy in twisting jets from the galactic core. Image courtesy of NASA.

Or to put it more broadly, when the galactic center is directly overhead and your body is most precisely aligned with the galactic core, your intuition spikes.

This effect is pretty much, if not exactly, as Sri Yukteswar described his concept of the Vishnu Nabhi. When we are in the moment of this Golden Alignment with the Vishnu Nabhi, the source of all Dharma and life energy, we are able to see things more clearly, to connect to God more easily, to, as Yukteswar put it, "*easily comprehend all, even the mysteries of Spirit.*"

So what we have here is no less than three interconnected cycles of time that govern our consciousness through the physics of the stars. We have a 26-thousand-year local Yuga cycle, probably connected to an elliptical orbit around an undiscovered binary companion to the Sun, a 26-million-year cycle mega-Yuga cycle, and a 260-million-year ultra-Yuga cycle linked to the Vishnu Nabhi at the center of our galaxy.

The questions at hand are: What do these Yuga cycles say about our future, and when can we expect to come out of the dregs of the Kali Yuga and move into the better world we all want? Not surprisingly, the answer to that is in the Sanskrit texts as well.

In one of the ancient texts called the *Brahma Vaivarta Purana*, the end of the Kali Yuga cycle is discussed:

> "When flowers will be begot within flowers, and fruits within fruits, then will the Yuga come to an end. And the clouds will pour rain unseasonably when the end of the Yuga approaches."

The exact meaning of this text is unclear, at least to me, but keep the part about the clouds pouring rain unseasonably in mind as we move through the next few chapters.

Other texts describe the end of the Kali Yuga cycle involving a battle between Kali, a demonic entity sired by the Shiva the

Destroyer, and Sri Kalki, the 10th and final avatar of the god Vishnu the Preserver. According to another ancient text, the *Vishnu Purana*, Kali is the source of all evil, corruption, and destruction in the world, pretty much the equivalent of the biblical Satan. Near the end of the Kali Yuga, Sri Kalki will return riding on a white horse to do battle with Kali and his dark forces. The world will then suffer a fiery end that will cleanse it of all evil, and a new golden age, the Satya Yuga, will begin.

Now, the interesting thing about this version of the end of the Kali Yuga is that we skip right over the ascending Bronze and Silver ages, and enter a special 10,000-year period of enlightenment while still in the midst of the Kali Yuga (this is based on the traditional 432,000 year estimate of the length of the Kali Yuga). So we jump from the Iron Age to the Golden Age in one swift blink of the eye of time. To me, this speaks of a very special time, a time when all of the conditions, including our evolving consciousness and the physics as defined by the energies of the planets, must push us into a period of dramatic (but not necessarily *traumatic*) upheaval.

And as we'll see, there is another learned ancient culture that seems to have tapped into a fundamental knowledge of that moment in future history. And they seem to think it is coming up very soon.

8 The Mayan Calendar

What we commonly refer to today as "The Mayan Calendar" is actually a series of distinct calendars and almanacs developed by several early Mesoamerican cultures that pre-dated the Maya. The Mayan calendar has many parallels with the calendars of earlier civilizations like the Olmecs and Zapotecs, but the Mayan version is far more complex and sophisticated by comparison. The Mayan's actually had numerous calendars—some say as many as 17—but they had two that were of primary importance: a traditional 365-day civil calendar called the *Haab*, and a 260-day sacred ritual calendar called the *Tzolkin*. The Tzolkin is the one we commonly associate with the term "Mayan Calendar."

The Dresden Codex

Most all of what we know about the Tzolkin ritual calendar is derived from a manuscript called the *Dresden Codex*, a paper document that is thought to have been created in the 1200s but that is also thought to have been copied from a much earlier Codex. The Dresden Codex is one of four surviving Mayan manuscripts, each of which is named for the city in which it resided when it was revealed to the world. The four surviving codices are the Paris Codex, the Madrid Codex, the Dresden Codex, and Grolier Codex.

Image 8.1. The Dresden Codex. Image from Wikimedia Commons.

Of these, the Dresden Codex is by far the most interesting. It contains a wealth of information on Mayan culture, rituals, science, astronomy, and mathematics, without which the calendar could probably not have been deciphered. It is written (painted, actually) on a single long sheet of paper reinforced by stucco and folded into an accordion shape, much like a modern-day road map. The Dresden Codex first appeared in 1739 when it was purchased by Johann Christian Götze, the Director of the Royal Library at Dresden. How and who he obtained the document from are shrouded in mystery, naturally, but its authenticity has never been questioned.

Nor is it debatable that the Codex shows the Maya to be a highly evolved people, at least in terms of their astronomical and scientific knowledge. The Dresden Codex is filled with astronomical charts of truly amazing accuracy, including charts of lunar eclipse cycles and cycles of Venus transits (occasions when Venus passes between the Earth and the Sun). It also contains detailed almanacs, astronomical and astrological tables, and religious references, including the Mayan creation myths. Contrary to the Spanish depiction of the Maya as bloodthirsty sun-god worshipers who practiced human sacrifice, the Codex shows them, at least in their earliest years, as sophisticated astronomers obsessed with the night sky.

The story of just how the Dresden Codex got to the West is almost as intriguing as what it contains. It is commonly accepted by scholars that when Cortez and the Spanish landed on the Yucatan peninsula in what is now modern-day Mexico, they discovered reams of paper documents like the Dresden Codex covering all aspects of Mayan history, civil activities, and religious and astronomical information. Supposedly on orders of the Vatican, Diego de Landa, a friar bent on replacing indigenous beliefs with the Christian faith, ordered all 1,562 of these codices destroyed. He wrote of it to the Vatican in 1566: "*We found a great number of books and since they contained nothing but superstitions and falsehoods of the devil we burned them, which they took most grievously, and which gave them great pain.*"

Yet somehow, the Dresden Codex survived, possibly because Cortez himself recognized its importance and had it sent back to Spain. What we know for sure is that, without it, no one would ever have been able to decode the Mayan calendar system.

The calendar itself is a wonder of the ancient world. The Mayans tracked a series of counts using a base 20 numbering system on a so-called calendar round, a stone device used to calculate days, months, and years. In the Mayan system, a day was called a *kin*, a 20-day month was called a *winal*, and a 360-day year was called *tun*. Twenty *tuns* made up a *k'atun*, and 20 *k'atuns* made up a single *b'ak'tun*. Using this and other charts from the Dresden Codex, the Maya were able to calculate lunar eclipses and the orbital cycles of Mars and Venus with an even greater accuracy than we can today.

The calendar is actually a set of interlocking wheels depicted on the calendar round, each counting at a slightly different pace, with the inner wheels driving the outer wheels much like the gears in a modern racing car's gearbox. It helps to think of the Mayan calendar round as a set of interconnected temporal gears, a gear box of time perhaps, with each inner circle clicking along and driving the bigger circles along bit by bit. In the case of the Maya, however, the purpose was not to deliver kinetic energy to the rear wheels but to track the passing of days, months, and years.

Image 8.2. The surviving Aztec Calendar Stone, a later version of the Mayan Calendar round. Image from Wikimedia Commons.

Biblical scholars have found it interesting through the years that the *b'ak'tun* cycle of 394.3 years also equates to 144,000 days, a number that has significance in the biblical numerology of the book of Revelation. In fact, the number 144,000 is mentioned three times in *Revelation*. Each of these verses tells the story of events that unfold at the end of time, when the good and the godly face down the evil and godless. Whether they are connected to something magical about the number itself is debatable, but the connection is interesting and there does seem to be a historical resonance for the number 144,000. It is also interesting that the Maya used a 360-day year instead of a 365-day year. In order to correct for this, they included five "dark days" at the end of the year to make the calendar match the Earth's true orbital period. I have always suspected that the 360 degrees of a circle are connected to this ancient Mesoamerican year of 360 days, because I believe that the original orbit of the planet Earth was 360 days, not the current 365 days.

In any event, the Maya then added up 13 (13 was a significant prime number for them) of the 394 year *b'ak'tuns* to come up with a total of 5,125 years, which is called the "long count" calendar cycle. So with the Mayan system, dates actually appear like this: 12.19.19.19.19. The number 12 is the *b'ak'tun* (there are 13 *b'ak'tun* cycles in a Long Count) and the rest of the numbers make up *k'atuns, tuns, winals,* and *kins.*

b'ak'tun	k'atun	tun	winal	kin
12	19	19	19	11

Depending on whom you ask, the current long count calendar cycle began on either August 13, 3114 BC (Gregorian), August 11, 3114 BC (Gregorian) or October 15, 3374 BC (Gregorian). As it is now understood, the long count cycle begins by turning the cyclical clock over to 13.0.0.0.0. In our Western way of thinking, it should logically begin at 0.0.0.0.0, but the Maya counted from one to 13 instead of zero to 12, so in their system 13.0.0.0.0 is essentially the "zero" date. Because the Maya tend to view time as cyclical rather than linear as we do in modern times, once the whole thing gets to 12.19.19.19.19, it clicks over again to 13.0.0.0.0 like the odometer in a car.

Again, depending on where you start, the end date may or may not come out on the 2012 winter solstice. But if you start with the now universally accepted August 11, 3114 BC date, the Long Count calendar cycle ends on December 21, 2012, which exactly correlates to the 2012 winter solstice. In general and with one notable exception, which we'll get to later, most everyone now agrees that the current 13 *b'ak'tun* Long Count cycle ends on December 21, 2012 (at 11:11 a.m. Greenwich Mean Time, to be exact about it) and that it began on August 11, 3114 BC. A new breed of scholars, like John Major Jenkins (*Maya Cosmogenesis*), argue that, because the end date falls on the winter Solstice of 2012, it cannot be a coincidence.[1] As they see it, the Maya were certainly advanced enough in their knowledge of the movements of the stars to know that particular date was the solstice.

The 3114 BC date is also interesting because the Maya identify it as the start of the so-called "Fourth World" of Man. According to the accounts in their creation myth texts, called the *Popol Vuh*, God has tried his hand at three previous versions of Man. Each World has lasted one 13 *b'ak'tun* calendar cycle, and, though each has met a different fate, they have each been destroyed in a swift global cataclysm. The *Popol Vuh* states that the start of the current world was the now established 3114 BC date, when the "Birth of Venus" took place. There has never been a correlation made between the August 11, 3114 BC date and any particularly significant astronomical events involving Venus, so it is at least possible that the Dresden Codex actually started from the *end* of the current 13 *b'ak'tun* cycle in 2012 and counted *backward* to get the 3114 BC start date. It certainly would be easier to have worked back from a known astronomical event rather than forward from essentially nothing, but I think it is significant that a series of early archeological finds seem to date from around this era. The Egyptian First dynasty, for instance, the dawn of true civilization in Egypt, is dated to around 3100 BC—which is also given as the start date for the current Hindu Kali Yuga cycle.

I find it more than a little intriguing all of these cycles seem to begin around this period form 3114 BC to 3100 BC. If the Mayans are correct in finding something significant to the breaking up of precessional cycles into 5,125-year chunks of time, then we really do have a convergence of thought from both cultures that this period in our history is important. Adding 5,125 years to the 3114 BC start date gets us to 2012. Adding the same number of years to the Yuga cycle start date of 3102 BC gets us to 2023. Perhaps we should worry more about the "2012 era," 2012–2023, than we do about any specific date.

So what then did the Maya mean by calling the August 11, 3114 BC date the "Birth of Venus?" We do know that the Maya were very concerned with the motions of Venus around the Earth, and there were a pair of Venus transits (Venus passing between the Earth and the Sun) in this 3114–3100 BC period. Why exactly this was important will be made clear later, but there is little question that these events are the "Birth of Venus" described in the codex. What is more important, however, is what the Dresden Codex was marking at the End of Time—December 21, 2012.

The Mayan Prophecies

These days it is common to hear stories about "the Mayan prophecies," on news channels and in various magazine articles and books. In reality, the Dresden Codex didn't contain anything that could properly be called a "prophecy," at least as we define it today. There are various researchers who have tried to twist the data to fit pet theories, but most of these fall apart under the stress of time, interpretation, and mathematics. Even some of the best researchers and spiritual scholars are not immune from getting caught up in the idea that the codex contained some specific predictions about the future.

Among these are Englishman Adrian Gilbert and Maurice Cotterell, the former a well-respected author and expert on the symbols of ancient Egypt, and the latter a well-known Mayan scholar and ancient mysteries enthusiast. In 1995, they released a book called (of all things) *The Mayan Prophecies* that sought to argue a mathematical correlation between the Dresden Codex and sunspot cycles. They concluded that when the current Long Count cycle ended, the Sun's output would rapidly increase and force a sudden reversal of Earth's magnetic poles. This would, naturally, be a really bad thing for all of us, causing storms, high winds, and rains of enormous proportions.

According to Gilbert and Cotterell, the "Seven Prophecies of the Maya," are a set of predictions that relate directly to our modern times. These prophesies are generally linked to the later Hopi prophecies, which we will discuss in the next chapter, and are said to tell of the events that will lead to sun's explosive outburst and the flipping of the magnetic poles.

The Seven Prophecies of the Maya are, in order:

1. Seven years after the beginning of the last K'atun (1992) a period of tribulation (1999–2012) would begin. This would shake the underpinnings of society and government, and lead to a rising consciousness that would be able to deal with the fear of change we were experiencing. Each of us, it is said, will face a "Hall of Mirrors" and confront our own behaviors and choices. It calls this period "the Time without time."
2. The increasing energy from the center of the galaxy will cause greater extremes and social polarization. Those that are already more spiritual will tune to this higher frequency, while those that are of lower vibration will react more emotionally and with great judgment of others, leading to more conflict in the world and contributing to the instability predicted in the first prophecy.

3. The climate will get much hotter, leading to hurricanes, alternating deluge and drought, crop failures, and famine. It blames this on a combination of increased solar output and man's inability to live in harmony with nature.
4. The polar ice caps will melt, reacting to the increased energy from the Sun especially, but also aided by our own selfish and environmentally abusive activities. It also talks of rising sea levels.
5. Fear and scarcity based monetary systems will collapse, brought on by our own awakening consciousness and a dependence on valueless (paper) currency. A new, single spiritual path will emerge for all of mankind.
6. During the period of tribulation (1999–2012) a great blue comet will appear as a harbinger and warning that the time has come to turn away from material desires and back to God.
7. At the End of Time, all of us will have a chance to turn inward to God and spirit, and that those that do will transition to a higher existence more attuned to a spiritual life.

In looking at these, I'd have to say that the Maya are at best 4-for-6, with number-7 still to be decided. It is hard to argue that the post-1999 period has been a difficult one, although it is a stretch to call it a "tribulation," at least in the biblical sense. It is true that in 1999 we were in a period of relative peace and money was plentiful. Since then, we've had the 2000 recession, 9/11, the Iraq and Afghanistan wars, the bursting of the housing bubble, and the 2007 financial crisis. But to those who remember the Jimmy Carter years and the 1978–1982 recession, things aren't much worse than they were then.

As to prophecy number two, I'd have to agree that our political discourse is getting far more polarized and partisan, with plenty of hate to go around. I also find it interesting that the Maya (at least according to some interpretations) believed that the center of the galaxy has an energetic influence on consciousness in the same way that Hindus did, yet these cultures were separated by thousands of miles and a couple of oceans, and had no contact. The third prophecy regarding rising temperatures, hurricanes, deluge, and drought; seems like a miss to me. Instead, the Sun has gone very quiet, temperatures have dropped for almost 10 straight years now, and there are no widespread crop failures or famines.

The fourth prophecy, speaking of melting polar ice caps and rising sea levels, is also clearly false. Despite hysterical claims of both occurring, the evidence is underwhelming at best, and recently a group of climate alarmists was caught conspiring to forge data to support their global warming claims. After several tough winters, no one believes the global warming crowd anymore, especially those hawking the man-made variety.

The fifth prophecy regarding the collapse of monetary structures also seems unlikely to be validated, although it is certainly too early to tell. As I write this in the spring of 2010, it is not clear what a summer of upheaval or the next election may bring.

The sixth prophecy, about a great blue comet coming to mark the era, is clearly true. On October 26, 2007, Comet 17P/Holmes passed through the solar system and suddenly exploded in a great blue fireball that was visible to the naked eye. At one point, its coma was 70 percent the size of the Sun, and within a few days it had expanded to become the largest body in the entire solar system. No one yet knows what caused the comet to explode as it did, and such a phenomenon has never before been observed by anyone in human history.

What this all means is debatable, but there is one claim that I have to call a "prophecy" that is *not* debatable. As I mentioned before, the Mayan origin myths revealed in the *Popol Vuh* describe a conflict over control of the Earth. In the story, the Earth is under the influence of an evil parrot God named Seven Macaw, who is deceiving the people of the Earth by claiming to be the Sun or the Moon. The Gods send a pair of hero twins, *Hunahpu* and *Xbalanque*, who are commonly accepted to be representative of the Sun and Venus. Hunahpu and Xbalanque do battle with Seven Macaw and vanquish him, but with the warning that he will return later and re-engage the battle for the Earth. The last page of the Dresden Codex shows an image of the resurrected Seven Macaw being aided by the god Kukulkan in unleashing a cosmic monster, a serpent god that is pouring water down upon the Earth from the sky.

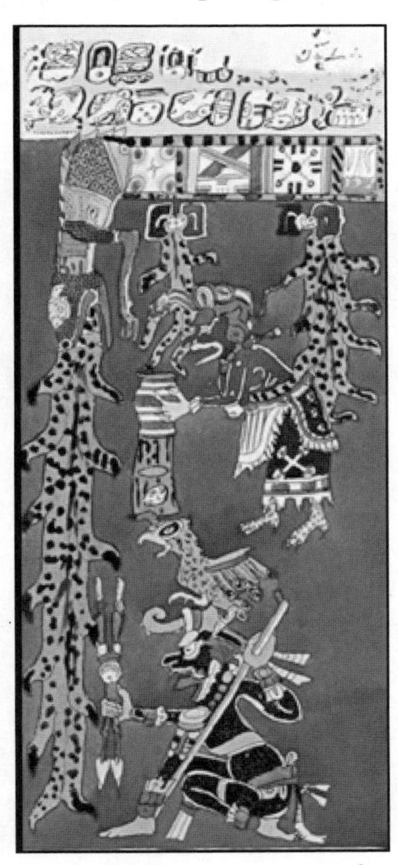

Image 8.3. Image courtesy of Sean David Morton.

Interestingly, if in fact Gilbert and Cotterell's scenario is correct and there is a massive solar flare sometime around 2012 that causes a sudden magnetic pole shift, this is exactly what could happen. Such a sudden realignment of the magnetic pole would cause high winds, terrible thunderstorms, and torrents and torrents of rain. Just as if water was pouring down on us from above.

Okay, ready for some good news?

There is an entirely different view of the Dresden Codex, the calendar, and the end of the long count cycle presented by a set of equally skilled scholars. And what they have had to say may go a long way to explaining why only a few of the Mayan end time prophecies have come true so far.

9 The Mayan Factor

For every calendar researcher who argues that the Mayan calendar spells out a portent of doom, there is another who argues that it is nothing of the kind. They see it as a spiritual blueprint, a road map of mankind's ascension to a higher way of thinking. Jose Arguelles, John Major Jenkins, and Carl Johann Calleman are three such men who have long led the charge against the End of Days alarmists.

Few people these days remember the original Harmonic Convergence in August, 1987. At the time, it was mostly ridiculed or ignored by a media monopoly that consisted of a few television networks and newspaper conglomerates. In this stuffy, controlled environment, author Jose Aguelles declared August 16th and 17th of 1987 to be a "Harmonic Convergence": a period of unique astronomical and astrological alignments that he claimed would transform the planet and usher in a new era of peace and cooperation. Despite the resistance from the state run media, the idea took hold and made an impact on the world.

In his book *The Mayan Factor*, Arguelles broke down the Dresden Codex using a complex and unique method of linkages to the Great Pyramid at Giza, the Mayan calendar, and other sources to map out the future history of the world. He called this system the "Dreamspell Count," and he concluded that when the current

Long Count calendar cycle ended on December 21, 2012, the world would in fact face a series of physical cataclysms that would rock it to its core and result in widespread devastation of human civilization. But hopefully, he also concluded that the world would have a chance to avert this devastation using the power of conscious, spiritual thought.

Aguelles argued that if just 144,000 souls (there's that number again) would come together in peace and love and harmony on the days of the convergence, then all the dire prophecies of the Dresden Codex would be diminished downward by a factor of 10. Rather than a mere 144,000 people coming together, millions gathered in virtually every free country of the world to celebrate, pray, and meditate on their desire for a more peaceful, harmonic world based on God's natural laws. The result was a dramatic drop in violence the world over. The Soviet empire fell, East and West Germany were reunited, and, other than the first Gulf War, there were no open armed conflicts between nations. Even the Gulf War itself was almost miraculous for its minimal loss of life, both civilian and military, compared to other conflicts.

In any event, not everyone in the emerging New Age community necessarily agreed with Aguelles, "Dreamspell Count." Even the Hopi elders, the far distant relatives and heirs to the Mayan culture, pointed out that, despite the demonstrably positive effects of the first Harmonic Convergence, the Mayans themselves never used Aguelles's system. Others, like Swedish researcher Carl Johan Calleman, have devised their own systems of calculating both the Mayan calendar end dates and that of the later Aztec calendar cycles.

Calleman has been studying the Mayan calendar since 1993. His conclusions, laid out in a series of books published in the last decade or so, are quite different yet strangely convergent with Aguelles's view of the future. Aguelles agrees with the interpretations of the Dresden Codex and other monuments that argue for some kind of cataclysmic events around 2012, but argues that the codex also provides for an "escape hatch" to avert the disaster scenario through the power of our own consciousness.

Calleman, conversely, doesn't buy into the physical catastrophe scenarios at all. He sees the Mayan calendar as a map to the spiritual development of human beings, a way of tracking our cycles of evolution through the eons. He predicts, based on his understanding of the calendar and his decoding of certain Mayan monuments, that the end of the current 13 *b'ak'tun* Mayan calendar cycle will manifest in the convergence of nine different spiritual entities into the physical world—and that our planet and our very existence will never be the same. These nine different spiritual entities are undeniably related to the significance of the number 9 in Mayan mythological lore.

Calleman pulled this idea primarily from the decoding of an inscription on Monument six in Tortuguero, Mexico, about 50 kilometers west of Palenque, where the nine-level Temple of the Inscriptions dominates the ceremonial center of the ancient Mayan city. Monument six contains one of the few specific surviving references to 2012, or the end date of the current Long Count cycle. The specific inscription refers to the "thirteenth Bak'tun," and a mysterious entity named *Bolon Yookte*.

Although it is damaged, the inscription has been at least partially deciphered:

> "The Thirteenth 'Bak'tun will be finished (on) Four Ajaw (2012), the Third of Uniiw (K'ank'in) will occur. (It will be) the descent of the Nine Support? God(s) to the...."

What exactly this inscription means is naturally the subject of much debate, but Calleman believes that the *Bolon Yookte* (nine support Gods) is both an entity and a measure of time that coincides with the end of the nine levels of spiritual development he sees predicted in all of the nine-level pyramids scattered throughout Mexico. What exactly the "descent" of Bolon Yookte means is far from clear, but hopefully it doesn't mean a downpour of water or one of the other catastrophe scenarios described earlier. How this connects to Mayan creation myths is interesting to say the least.

As I've already established, the number 13 had a great significance to the early Maya, as did the number 9. The Maya viewed the universe as consisting of Heavens above and Underworlds below, with the human world existing in between. The Heavens consisted of 13 layers stacked above the Earth, and the Earth rested on the back of a turtle floating in the cosmic ocean. Below the Earth lay a realm called *Xibalba*, an Underworld that existed in nine levels or layers. Uniting these three realms was a giant Tree of Life called *Yaxche,* whose roots reached into the Underworld and whose branches stretched to Heaven. The Gods and the souls of the dead traveled between the worlds along this tree. There is an interesting connection between the tree of life, rooted in the depths of the Earth and extending to heaven, and the Egyptian Djed pillar. Both could be described as axial supports for reality itself, holding up the heavens and the Earth so that creation might continue, and both are arguably references to the rotational axis of the Earth.

According to Calleman, the Mayans believed the nine different realms of the Underworld existed timelessly, but humanity experiences them as successive ages or epochs. We are gradually working our way through them to the ultimate goal: the transition to the final stage, or Universal Underworld, which will reunite the human race in co-creation with God. In this way the nine Underworlds represented not just levels of conscious development or physical spaces, but periods of time as well.

This fundamental idea has been expressed by the Maya not just in their calendars and codices, but also in their monumental architecture. Each of the major cultural centers of the ancient Mayan world has a nine-level pyramid as its central structure. Whether it be the Pyramid of the Plumed Serpent in Chichen-Itza, the Pyramid of the Jaguar in Tikal, or the Temple of the Inscriptions in Palenque, the significance of the nine levels of evolution and consciousness is impossible to ignore, as it is relentlessly embedded in stone.

These nine Underworlds each span a specific period of linear time, each successive Underworld being shorter by a significant degree than the previous one, and each representing a specific type of physical or spiritual manifestation. The nine Underworlds are, from bottom to top, as depicted in the Mayan pyramids: the Cellular Underworld, the Mammalian Underworld, the Familial Underworld, the Tribal Underworld, the Regional Underworld, the National Underworld, the Planetary Underworld, the Galactic Underworld, and finally the Universal Underworld. It is this last phase, the achievement of a universal consciousness united with God, that Calleman sees as the attainment of the *Bolon Yookte* state of mind and spirit. So the descent of the Bolon Yookte is the unification of all of the previous levels of development in one new and whole consciousness.

Each Underworld is based on the *tun*, the Mayan 360-day year (which, as I said earlier, I suspect is the original and correct orbital period of the Earth around the Sun). Each Underworld also has 13 creation cycles (called "Heavens") within it in which to accomplish its required task. In the case of each these creation cycles, the number of days is multiplied by 13 to find the corresponding duration of each Underworld. Using this system, the first Heaven of the first Underworld has lasted some 460,800,000,000 (460 billion, 800 million) days, or *kin*, as the Maya called them. This time period, called a *hablatun* by the Maya, equates to 1.26 billion of our sidereal years. This 1.26-billion-year cycle is then multiplied by the 13 Heavens (cycles of this creation) to get the number 5,990,400,000,000 (5 trillion, 990 billion, 400 million days) or 16.4 billion years. By this method, the Maya calculate that the universe is therefore 16.4 billion years old.

During this first Underworld matter was formed and the earliest forms of life-consciousness, cellular life, physically manifested on Earth and presumably throughout the rest of our universe. Thus the first Underworld forms the foundation for all others to follow, which is why it is represented by the first and widest of the pyramid steps. As the foundation, this period of creation is still going on, for if it ceased none of the other subsequent higher levels could be sustained.

Just as the Mayan pyramids step upward at an angle, resulting in ever smaller levels, Calleman also sees each of the Underworlds getting progressively shorter. By Calleman's calculations, we are currently in the next-to-last Underworld, the Galactic Underworld, where we will begin to view ourselves in the context of the galaxy around us, rather than from a planetary perspective. The current Galactic Underworld cycle began in May of 1999, and will, according to Calleman, run its course by 2011.

This is one of the major ways that Calleman differs from other Mayan scholars. Because of his calculation of the time spans of the nine Underworld's, he argues that the end date of the calendar is not the canonical December 21, 2012 date as recited by so many others, like John Major Jenkins, but rather the less-well-known date of October 28, 2011. In speaking of the October 28, 2011 date on his Website, Calleman contends:

> ...this date emanates organically from an evolutionary process and is not associated with any real or purported physical or astronomical event. It is simply the day when the universe, after nine major quantum steps, starting with the beginning of the universe, attains its highest energy state and there is no logical reason that this would mean the end of the world. It is just an evolutionary completion point when Bolon Yookte fully descends.

In other words, "the descent of Bolon Yookte" is the achievement of a blissful and peak state of consciousness, and nothing more. However, if the state of *Bolon Yookte* is the beginning of a completely new experience for humankind, who is to say that *both* dates aren't correct? It simply comes down to the age old question of when life begins: at birth—or conception? If we view Calleman's more Aztec-influenced 2011 end date for the Long Count cycle as the conception, and the more popularly known 2012 date as the birth of this new stage in our evolution, then there really is no conflict between the dates, is there?

Beyond this, Calleman also differs from other Mayan scholars in several other significant ways. Using his own interpretation of the length of each Underworld epoch, he then broke them down into a wave pattern of 13 phases, with the positive upswings being known as Days and the negative aspects of the wave being seen as Nights. According to Calleman's mapping of these day and night cycles, the 13 phases would be the First Day (stage 1), followed by the First Night (stage 2), and so on. If there are in fact 13 such phases of each Underworld, then each Underworld should end on the 7th day.

"And on the 7th day, he rested...."

If Calleman's process is sound, we are currently in the Sixth Night of the wave movement known as the Galactic Underworld, the next-to-last phase of our conscious development as human beings. By his reckoning, we will come out of the Sixth Night on November 2, 2010, about the time this book comes out, and if he's right we will then ascend back up the curve and ride the wave to the Seventh Day, which again he views as October 28, 2011. He sees the Seventh Day as the end of Cycles and the attainment of what the Maya viewed as the Universal Consciousness, the state of co-creation with God.

Calleman's perspective is that, as a road map to the development of human consciousness, the calendar could be used to predict cycles of human behavior. His economic predictions, based on this reading of the Calendar, have been eerily prophetic.

In his book *The Mayan Calendar and the Transformation of Consciousness*, published in 2004, Calleman wrote: "*Regardless of what forms such a [financial] collapse may take, it seems that the best bet is for it to occur close to the time that the Fifth NIGHT begins, in November 2007 [strictly speaking the 19th]*" (page 233). It is now generally agreed that the "housing bubble" burst in late November or early December 2007, precipitating a dramatic drop in the dollar and eventually imperiling the entire United States economy. This prediction was originally referenced in more general terms in an earlier Calleman book, *Solving the Greatest Mystery of Our Time: The Mayan Calendar*, written in 1999 and published in 2001 (page 187).

At the moment I write these words, we are in the Sixth Night of the Galactic Underworld, which began at midnight on November 8, 2009. I myself had a profound spiritual experience within an hour of that moment, and when I realized later that it was linked to the coming of the Sixth Night, it truly reinforced my confidence in Calleman's methods. It was only a few weeks later, when I first heard of Calleman and the so-called Sixth Night from my friend Sean David Morton, that I realized how intensely my heart had been transformed by that experience. I understood, in the only way that we humans truly can—by experiencing it myself—that there is a wind of change coming. And it is coming whether we want it to or not.

This is not to say, however, that Calleman's predictions, however impressive they may be, are the only Truth to be learned here. I myself see several flaws in his thinking that need to be addressed.

His conclusions are based on what he views as the more correct reading of the calendar, rooted in the idea that human events are cycles laid out by nodes on the so-called "Tree of Life." He insists that the Maya/Aztec system is a calendar based not in the material aspects of astronomy, as all of today's modern calendars are, but rather one based in the timeless aspects of Spirit. Calleman's view is therefore that the 2012 end date promoters are looking only at the physical, not the spiritual, and as such they will never see the Truth of what will happen in the 2010–2012 period.

That said, he clearly sees the next few years as chaotic and uncomfortable at the very least, and it's evident he does not believe that everyone currently on Earth will physically make it through the transition. Still, he has little regard for the "alignment crowd," as he calls them, because he sees them as hopelessly rooted in a modernist linear time/astronomy, based way of thinking. But Calleman's perspective is just as flawed, for he in his advocacy also ignores the influence of the astronomical in the bigger picture. Because he is most likely constrained by the limitations of the quantum physics theorists, he fails to see another possibility: that both the physical and spiritual are linked, if not inseparable.

In my model, it's the *physics* that drives the *consciousness*, it's the *science* that drives the *spirit*; it is the *astronomy* that drives the *astrology*. What this means is that we need to accept that there is a science behind our thoughts, moods, and actions, and that means we can counteract any negative energy that they generate in the midst of our coming existential crisis.

In our own sometimes smugly superior way, those of us who have rediscovered our connection with God can easily overlook or dismiss the left-brained, intellectual thought processes that sometimes keep us from truly connecting to the inner Divine. I find myself doing it all the time lately; it's a hard habit to break. But I strongly advise you all not to do this. The human body is an amazing device, and there must be a reason why that same God gave us the capacity to both think and feel. I will not delude myself into believing I know why that is. I merely caution you not to polarize either your mind or your heart. In my own journey, this is what I have tried to do, and as a result of not taking sides, I think I have had a glimpse of something that may be very important for all of us to consider:

What if neither side, the October 2011 or the December 2012 advocates, is wrong? What if neither side, those that argue that the Mayan calendar predicts only a spiritual transformation, nor those that insist we are facing a traumatic physical ordeal, is wrong? What if they are both right? What if time itself is both spiritual *and* physical, and what then must we do if there are indeed changes coming in both of these aspects of our existence?

What we have to do, in my view, is make the Choice.

10 The Hopi Prophecies

In 1992, a group of Hopi elders visited the United Nations on a mission of peace and hope. It was the fourth time since 1948 that they had visited the United Nations seeking an audience for what the elders called the "first people." The first three times, they were rejected. The fourth time, their proposal was accepted and they were granted an audience on December 10, 1992. That date is exactly one *k'atun* (20 years) and 11 *kin* (days) from December 21, 2012 (the significance of the number 11 will become clear a bit later). This interesting Mayan connection is made even more curious by the fact that the Hopi consider themselves to be the direct lineal descendants of the Mayans, the Aztecs, and Anasazi people of the four corners region of the American southwest.

The delegation was led by a man nicknamed Grandfather Thomas Banyacya (pronounced like "Bianca"), an elder and leader of the Hopi tribe. According to Banyacya, the leaders of the tribe had met in 1948 to discuss "things I felt strongly were of great importance to all people." He was told at that meeting that someday, the leaders of the world would assemble in a "great house of mica," a Hopi word that translates as "house of glass." In this great house of glass, he was told, men would meet under rules and regulations to solve the world problems without war. I cannot think of a better description of the UN building in New York City, where world leaders meet (ostensibly) to solve the world's problems without war in a building that is layered in glass from top to bottom.

Banyacya was one of four leaders selected in 1948 to deliver the Hopi message, and by the time the audience was granted, he was the only one still living. The elders had told them to knock four times, and that when they were admitted, they were to deliver prophecies handed down from "the time the previous world was destroyed by flood and our ancestors came to this land." Banyacya was then given a sacred prayer feather, which he was to deliver to the leaders of the house of mica and to open the house to native or "first" people. During his short, 10-minute speech, he gave the assembled delegates a stern warning of the future based on Hopi prophecies he had been given at that 1948 meeting.

He began by introducing the members to the Hopi philosophy of hope, life, harmony with nature, and commitment to the spiritual teachings of Massau'u, the Great Spirit. Next, he informed them of the Hopi view of the current state of the world:

> At the meeting in 1948, Hopi leaders 80, 90, and even 100 years old explained that the creator (Great Spirit) made the first world in perfect balance where humans spoke one language, but humans turned away from moral and spiritual principles. They misused their spiritual powers for selfish purposes. They did not follow nature's rules. Eventually the world was destroyed by sinking of land and separation of land by what you would call major earthquakes. Many died and only a small handful survived.
>
> Then this handful of peaceful people came into the second world. They repeated their mistakes and the world was destroyed by a freezing which you call the great Ice Age.
>
> The few survivors entered the third world. That world lasted a long time and as in previous worlds, the people spoke one language. The people invented many machines and conveniences of high technology, some of which have not yet been seen in this

> age. They even had spiritual powers that they used for good. They gradually turned away from natural laws and pursued only material things and finally only gambled while they ridiculed spiritual principles. No one stopped them from this course and the world was destroyed by the great flood that many nations still recall in their ancient history or in their religions.
>
> The Elders said again only small groups escaped and came to this fourth world where we now live. Our world is in terrible shape again even though the Great Spirit gave us different languages and sent us to four corners of the world and told us to take care of the Earth and all that is in it.

Banyacya then held up a ceremonial rattle, which he said displayed a time line that was rapidly running out. He said that when the end of the time line was reached, the prophecies of the Hopi people would cease, and that this time was soon upon us. He did not recite the specific Hopi prophecies of the end times, only issuing a general warning.

The Hopi prophecies have never been officially published, but an account of them has emerged from an encounter by a minister named David Young. In the early 1960s Young stopped to pick up a hitchhiker. He was an American Indian man who later identified himself as a Hopi elder named White Feather. Young's account of what White Feather told him was eventually published in 1963 by Frank Waters in his book *The Book of the Hopi*. Obviously, I strongly suspect that White Feather is one of the three other Hopi who were entrusted with the message for the United Nations along with Thomas Banyacya.

According *to The Book of the Hopi*, the nine Hopi prophecies are:

> This is the First Sign: We are told of the coming of the white-skinned men, like Pahana, but not living like Pahana, men who took the land that was not theirs. And men who struck their enemies with thunder.

"Pahana," for the record, is a great white brother who it is said will come to the world in the end times and lead the way to purification of the Earth and re-introduce men to the philosophies of Great Spirit, the Indian God who exemplifies the Universe. The prophecy itself is not so remarkable: It simply tells of the coming of the selfish white man and of his weapons, like guns and cannons. Without any record as to when it was actually first revealed, it is hard to give it much credibility as a prophecy or prediction, and because the Hopi tradition is oral it was first written down long after the coming of the white man. The second prophecy falls into the same category:

> This is the Second Sign: Our lands will see the coming of spinning wheels filled with voices. In his youth, my father saw this prophecy come true with his eyes—the white men bringing their families in wagons across the prairies.

This is an obvious reference to the wagon trains that crossed the American West, but again there is no real way to verify it as authentic for the same reasons listed previously.

The third prophecy tells of the coming of the longhorn cattle:

> This is the Third Sign: A strange beast like a buffalo but with great long horns, will overrun the land in large numbers. These White Feather saw with his eyes—the coming of the white men's cattle.

The fourth prophecy appears to be related to the coming of the railroads:

> This is the Fourth Sign: The land will be crossed by snakes of iron.

The meaning of the fifth sign is less clear, but it is generally accepted as relating to the telegraph, telephone, and power lines that now connect the entire United States:

> This is the Fifth Sign: The land shall be criss-crossed by a giant spider's web.

The sixth prophecy, *"The land shall be criss-crossed with rivers of stone that make pictures in the sun,"* appears to be connected to the interstate highway system, and the manner in which it creates mirages in the distance. The seventh prophecy seems an obvious reference to oil spills, much like the one in the Gulf of Mexico in May 2010:

> This is the Seventh Sign: You will hear of the sea turning black, and many living things dying because of it.

The last two prophecies of White Feather are the most concurrent with modern times. The eighth seems to be a reference to the hippie culture of the 1960s, which at the time these were first published was still more than half a decade away:

> This is the Eighth Sign: You will see many youth, who wear their hair long like my people, come and join the tribal nations, to learn their ways and wisdom.

But where it really gets interesting and relevant to 2012 and our immediate future is in the ninth prophecy:

> And this is the Ninth and Last Sign: You will hear of a dwelling-place in the heavens, above the earth, that shall fall with a great crash. It will appear as a blue star. Very soon after this, the ceremonies of my people will cease.

There are those who consider this to be a reference to the same event in the so-called Mayan Prophecies listed in Chapter 9, which tell of a great blue comet falling from the sky as a harbinger. Certainly, as I mentioned in that chapter, Comet 17P/Holmes fits the description and also fits a version of the Hopi prophecies that speaks of the blue star called "Kachina." According to one version of the prophecy, "*When the Blue Star Kachina makes its appearance in the heavens, the Fifth World will emerge*." What this means is that the falling of the great blue star will signal the start of the great cleansing, when all those who have embraced materialism over spirit will be washed from the Earth in a great upheaval, and those few that are left will begin a new world, the Fifth world.

There are some who speak of the re-entry of the American Skylab space station or the space shuttle *Columbia* disaster as fulfilling the prophecy of the blue star Kachina, because both fell from the sky and appeared blue as they descended. But I am not so sure that this prophecy has been fulfilled, and, in fact, I think it has yet to happen.

The specific prophecy of White Feather doesn't just speak of a blue star; it speaks of a "dwelling place in the heavens" that shall *appear* as a blue star. Right now above our heads is the International Space Station, the only existing (or publically acknowledged) "dwelling place in the heavens" at the moment. Other Hopi translations connect the blue star Kachina with Sirius, and it is said that Kachina is literally the Hopi word for Sirius. However, it seems unlikely if not impossible that Sirius, which is a blue-colored star by the way, could fall from the night sky, or that it could be considered a dwelling place in the heavens. Except…

As we learned in Chapter 6, the Egyptians linked the star Sirius with the goddess Isis, literally the mother of creation and nurturer of Man. The International Space Station, at least the American part of it, also has an official name: *Freedom*. But the entire space station is most commonly referred to by its acronym, "ISS." Now, if you take the ISS acronym and say it as a word, you really have two pronunciation options. You can say it like a snake's hiss, "Issssss…," or you can insert an apostrophe, IS'S, and pronounce it "Isis."

If you have read *Dark Mission* and appreciate all of the NASA/Egyptian connections Richard and I outlined in that volume, you might just consider that this naming and pronunciation is intentional. Whether or not that's true, if the dwelling place in the heavens named for the blue star Kachina, the *IS'S*, were to fall from the sky, it would certainly qualify as fulfilling the ninth prophecy of White Feather. Fortunately, as I write this, it is still in the sky, orbiting comfortably above us.

There is also one other prophecy that has gained some acceptance among Hopi scholars. Not listed among the others in *The Book of Hopi*, it states that if man journeys to "sister moon," he should take great care not to bring anything back with him. If we do, the prophecy says, it will create a great imbalance in the Earth and we will suffer droughts, earthquakes, storms, and social unrest. Of course, we did journey to the Moon, and we did bring something back with us, probably far more than just rocks (see *Dark Mission*). Unquestionably all of the predictions of this prophecy have come true, but it is hard to say that any of these consequences are any more pronounced or frequent than they were before the Moon landings.

Okay, having said all this, before you start to get too worried, keep in I mind that the Hopi also promise that all of this can be avoided, right up until the last minute.

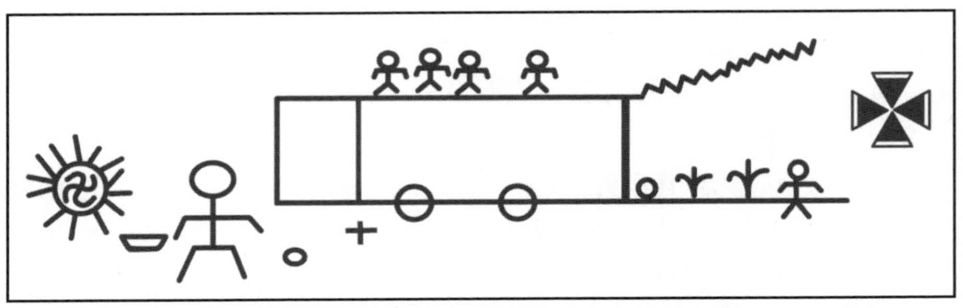

Image 10.1. Artistic depiction of Prophecy Rock as presented by Thomas Bayancya to the United Nations on December 10, 1992. Image by Krys Lilly.

One of the other items that were presented by Bayancya at the UN was a photograph and artistic depiction of something called Prophecy Rock in the Hopi lands in Arizona. Sitting high atop a mountain named Second Mesa in the four corners area of Arizona,

Prophecy Rock contains a petroglyph that is said to be a depiction of the end of the Fourth world and the beginning of the Fifth, and as such it tells the story of the future from the Hopi perspective. According to Bayancya and the *Book of the Hopi*, the interpretation of the figures and symbols is as follows:

The large human figure on the left is the Great Spirit. The bow in his right hand represents his instructions to the Hopi to lay down their weapons—to literally "beat their swords into plowshares." The swastika in the sun and the Celtic cross represent the two helpers of *Pahana*, the True White Brother who shall return at the end of the Fourth world to guide the Hopi to their new lands, where they will start human civilization over yet again. The swastika, for the record, is an ancient Hindu symbol that represents the Sun as a star gate, the source from which all life energy emanates from "higher places." It was appropriated by the regime of Nazi Germany but does not originate with them. The Celtic cross is also identical to the Hopi sun symbol, and represents the four corners of the world where the survivors of the third world emigrated.

The vertical line to the right of the Great Spirit is a time scale in thousands of years. The point at which the Great Spirit touches the line is the time of his return. The circle next to Great Spirit represents the cycle of time that has begun again.

To the right of the vertical time line are two horizontal paths, one above and one below. The upper path represents the path of intellect and technology, the lower line the path of spirit and a life of harmony with God and the Earth. The vertical line above the Christian cross represents the coming of the white man to native lands. On the upper path are four human figures, representing each of the previous three Worlds and the current one. Their heads are detached from their bodies, meaning that the path of intellect and materialism will invariably lead to the separation of the head from the heart. These "Two-Hearted" people have lost their way and turned away from God.

On the lower path are two circles intersecting the line, and this is said to represent two great shakings of the earth that will serve as warnings to the Two-Hearted people to mend their ways. It is commonly accepted today that these great shakings are World War I and World War II.

Note that the fourth figure on the upper path is separated from the other three, indicating that he still has a chance to choose another path. The second and last broad vertical line connecting the upper path with the lower path is the Choice point for the Two-Hearted people of the fourth world. It means that near the end of the Fourth World, whenever that is, people will have a Choice—a chance to turn away from the material and toward the light of God. If they do, they will find themselves on the lower, more harmonious path, the path of the One-Hearted people.

Very shortly after the Choice point has passed, we see that the upper path becomes jagged and then ends. This symbolically means that those who stay on the upper path will face chaos, upheaval, and eventually extinction. Those who choose to return to the lower path will see the third great shaking of the Earth, as represented by the third circle on the bottom path. This is the day of the great cleansing, the day when the Earth will be rid of all those who fail to heed Great Spirit's warnings. After that, the corn will grow abundantly and all men will become One-Hearted once again.

I know a lot of this is scary, but it is also filled with hope. Just like the Mayan prophecies and the Yuga cycles, this message says that not only can anyone be saved by turning inward in their hearts to God, so too can the entire world be saved by choosing to abandon the immoral and the material. There are numerous Hopi prophecy scholars who insist that the prophecies make it clear that this is a question of Choice—that the day of purification doesn't have to come at all. But, like the Maya and the Vedas, they do seem to be insisting that the day of Choice, the Day of Judgment, is coming soon, most probably in our lifetime. Only the Egyptians, who tracked precession by observations of Orion, would seem to disagree that we are close to the end, placing the end of the current cycle in 2175, well beyond most of our lifetimes.

So what we have here are four ancient cultures—the Egyptians, the Mayans, the Hindus, and the Hopi, who all agree that we seem to be nearing the end of a cycle. These cycles seem to be linked to precession or at least the movements of the heavens, and there does seem to be a reason to suspect that the 2012 end date of the Mayan calendar long count and end of the current Kali Yuga cycle are connected. In addition, each of these cultures speaks of the end of these cycles as involving some sort of physical Earth changes, possibly very big and very traumatic. What we really don't know yet is just what might happen, and how it might come about. It is all well and good to talk about pole shifts and solar flares and other natural disasters, but why should they happen now? What is so special about this period in time, and how is it connected to physics, the real physics, of this 2012 era?

11 Torsion

For most of the 20th century, physicists and mathematicians have sought what essentially amounts to the Holy Grail of physics, something called the "unified field theory." Ever since James Clerk Maxwell's early work unifying electricity and magnetism into "electromagnetism," it has been hoped that someone will find a way to unify the behavior of forces and particles under one specific "field," that is, a medium through which these force waves of energy and particles of matter can travel and interact and have it all make sense. Right now, because of all of the conflicts and issues with quantum physics and Newtonian mechanics and relativity, there is simply no theory of everything that can account for all that we can measure and observe in the Universe. God and his works remain elusive.

Even though we consider most of space to be a "vacuum," a region devoid of matter and energy, in reality most of that vacuum is occupied by at least a few sub-atomic particles or some kind of radiated energy. In order for this to be the case, there must be a field in which this matter and energy exist and propagate. Wikipedia defines a "field" in physics as "a physical quantity associated to each point of spacetime." In other words, it views physical reality as a series of points in space that have some numeric value, which is used to describe the field's behavior. In this way, the entire Universe can be thought of as a field containing matter and/or energy.

The Earth's gravitational field, for instance, exists all around us, keeping us firmly tethered to the ground, and at any given point within that field there is a measurement that can be taken that defines that gravitational field in some way. In the case of gravity, this means that a point in the center of my body would part of the Earth's gravitational field, and it would have both a direction and a "magnitude," a measure of the intensity of that field. On the surface of the Earth, the magnitude of the gravitational field is more than enough to keep me in place, but if I was in orbit around the Earth, say in the International Space Station, the influence of gravity would be so slight that I would be essentially weightless and float freely.

Another aspect of the known and accepted fields is that they are usually defined by the limits of their influence. For instance, the Earth's gravitational field extends beyond the orbit of the Moon, but then it gets gradually weaker until it ceases to have any effect. Where the unified field would differ is that it would be the background field that exists everywhere and connects everything—literally space itself and all the other fields in it, like the Earth's gravitational field, and it would have no limitations. It would be of the same intensity in all places in the universe and at all times. In this book, we have called this unified field the Aether, and argued that it exists in a higher dimensional plane. But mainstream physics says it has yet to be found in this physical 3D reality.

What all this adds up to is that according to the "laws" of physics, forces between objects must pass through something, some kind of intervening medium, to have an effect on each other. That medium is called a field. What the mainstream guys have been unable to resolve to this point is how this all works, because different forces and different force fields seem to operate differently and without harmony. The four known and accepted forces are the strong nuclear force (which holds the little stuff, like sub-atomic particles, together), the weak nuclear force, electromagnetism, and gravity.

So without an acceptable field theory to explain how all these forces interact, physics is left with many problems and paradoxes.

As I showed you in the earlier chapters on physics, the mainstream thinkers will never be able to resolve the built-in paradoxes in modern physics because they are lacking the crucial linking mathematical components: higher spatial dimensions. Because of Oliver Heaviside's now-infamous editing of Maxwell's equations, modern quantum physicists just can't wrap their heads around this crucial idea and how it works. To this day, even some of the most edgy mainstream thinkers (there's a contradiction) will dip their toes in the waters of hyper-dimensions, but call it by some acceptable pseudonym like the "quantum medium" to try and explain things they see in experiments. For all of their books on hyperspace and Quantum Time, guys like Nassim Haramein and Michiu Kaku are still stuck in their three dimensional mindset.

Even out-of-the-box thinkers, like David Wilcock, still refer to the "fluid Aether" as if it is some undiscovered aspect of three-dimensional space, like Einstein's curved fabric of space time. It isn't. The Aether and all the interactions take place in higher dimensions. But, as we've learned before, we can measure the effects of these connections and interactions here in three dimensions. And more than that, after years of thought and experiments which have generally been ignored by science and the media, we can finally identify the mechanism through which the effects can be measured, predicted, and put to practical use. It's called torsion.

The concept of the metric tensor-torsion field was first asserted in 1913 in a paper by Dr. Eli Cartan. Cartan went to work on Einstein's theory of general relativity because it could not explain a phenomenon called spin-orbit coupling. Put simply, atoms are made up of a nucleus that contains protons and neutrons, and these particles are orbited by a negatively charged particle called an electron. When something is spinning (as we've established *everything* is) but also in motion (that is, also moving in a specific direction) electrons are known to have a variable energy output. This is flatly impossible in general relativity, so Cartan had to come up with some way to normalize (fix) the equations so that this could be explained. In the hyper-dimensional model, this is easily accounted for because it is the spin itself that is generating the energy, and it will vary as the little atom interacts with other atoms. At any rate, Cartan's solution was to say that not only did Einstein's space-time *curve* (remember that?) but it also *twisted*. He called this motion *torsion*, and named the field a "metric tensor-torsion field." But, don't be misled by the description. Torsion is not a field. The *Aether* is the field, existing in higher dimensions. Torsion is just the mechanism, the signature of the energy being gated which proves the Aether exists. But in the mainstream thinking, everything has to have a three-dimensional "field" to pass through, so Cartan gave us the metric tensor-torsion field.

Theoretically, torsion is the mechanism by which higher dimensional energies such as those we've been talking about all along might propagate through space. Because of the distinctive twisting aspects of torsion, the signatures left behind should be equally distinctive, characterized by equally twisting, spinning, roiling movements of energy and matter. Mainstream science sources, like Wikipedia, dismiss torsion as a "pseudoscience," and claim that there is no evidence that torsion effects are real and that advocates of the idea are "frauds." The reality is far different. There is in fact all kinds of evidence that the Universe adheres to the rules of torsion physics.

There are numerous examples of torsion effects seen in various sources from any number of NASA instruments. In 1998, the SOHO spacecraft imaged a spectacular solar flare that was twisting and spinning like some enormous strand of stellar DNA. There is nothing in any of the mainstream physics texts that can explain this twisting motion, but it is built-in to the torsion concept.

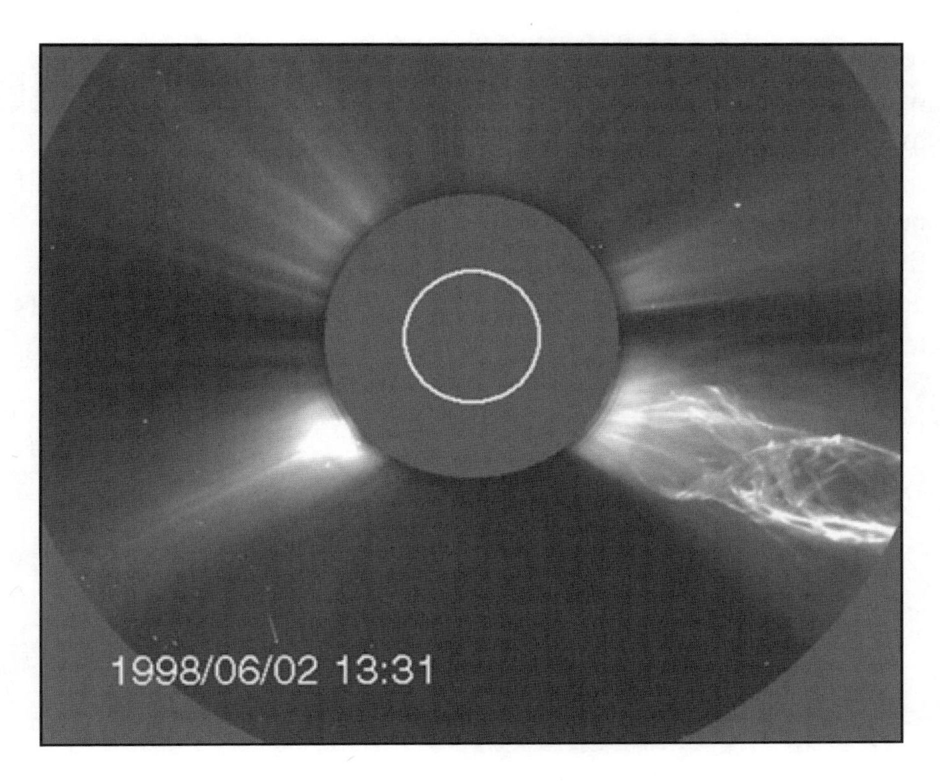

Image 11.1. SOHO image of twisting torsional solar coronal mass ejection. Image courtesy of NASA.

Another example is the Sun's motion around the galactic core. Viewed from above, the path is elliptical. Viewed edge on, the motion would appear to be an up and down flow, like a wave. But what is rarely mentioned beyond that is the fact that while the Sun is in orbit around the galactic center, it also twists and spins around the centerline of the orbital path in an unmistakable twisting, torsional pattern.

The principal objections to the concept of torsion as a unifying aspect of physics are that Cartan's original work concluded that torsion is simply too weak a force to have anything but the most minute effect on the Universe at large. His calculations were that torsional forces would be some 30 orders of magnitude weaker than gravity, which is already some 40 orders of magnitude weaker than electromagnetic energy. As a result, the Einstein-Cartan theory declared that torsion was a miniscule effect that was static (that is, it could not move or propagate of its own accord) and was therefore confined to a space smaller than the atom itself. However, in the 1970s, a number of research papers on the effects of torsion came to a different conclusion.

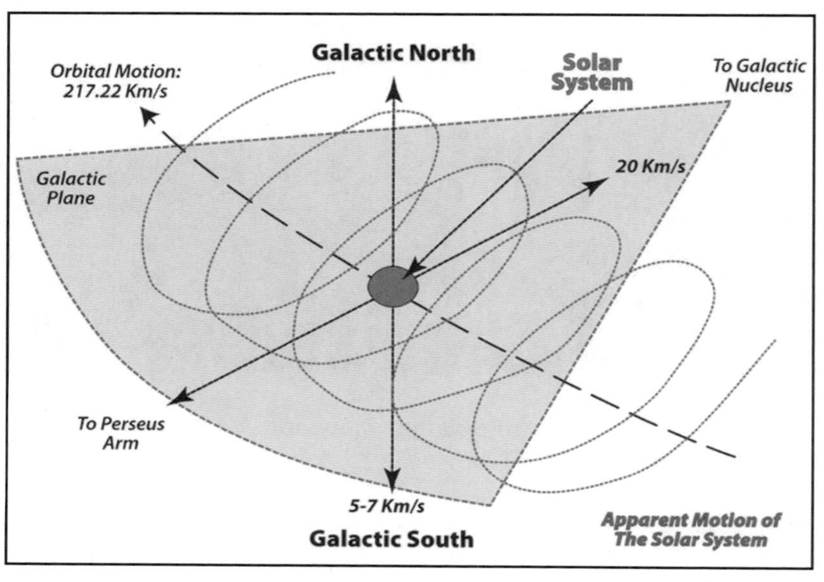

Image 11.2. The solar system's orbit through the galaxy adheres to all the concepts of torsion. As it flows along the elliptical orbit, it also rises and falls and twists and spins in a classic spin-orbit coupling relationship. Image by Krys Lilly.

A group of physicists and engineers, led primarily by British physicist Dennis William Sciama, showed that there were two distinct types of torsion effects: one that worked at the sub-atomic level (static) and one that worked at the planetary level (dynamic). Torsion that was a weak force and contained within the atom was confined to spinning particles (like the electron) that do not inherently radiate significant amounts of energy. But objects that were much larger, like the Sun, and that were in motion *and* spinning, generated what he termed "dynamic torsion fields." In other words, torsion had different properties on the large scale than it did at the small scale. The key to the dynamic torsion was the fact that there had to be two types of motion; like the Sun's spin combined with its orbital rotation around the galactic core. Under these dynamic conditions, not only could torsion produce a significant energy output, it could propagate that energy at super-luminal speeds—faster than the perceived speed limit of the Universe, the speed of light. This is why Dr. Bruce DePalma's little spinning ball experiment worked so elegantly. By spinning the ball to hyper velocities—27,000 RPM—he was able to open a gateway to a higher dimension and draw from it, at least enough to make the little ball fly higher and fall faster than the static ball once he added the second motion component, the ejection of the ball out of the test rig.

Other experiments also supported Sciama's conclusions. Dr. Hal Puthoff, a physicist who worked for the CIA and the National Security Agency, also conducted various experiments into ESP at the Stanford Research Institute and developed an interest in alternative energy theories. Along the way, he developed an interest in something called "zero-point energy," or energy from the vacuum. The basic principal of zero-point energy is that you get something from nothing, the nothing being the supposedly empty vacuum of space. In experiments conducted in the early 20th century, before the quantum mechanics theory was even proposed, scientists created mini vacuum chambers to simulate the vacuum of space. First, all of the air was removed from the chambers, and then the entire chamber was surrounded by lead shielding called a Faraday cage. The Faraday cage effectively shields the chamber inside from any static electrical fields, meaning that there is no electromagnetic energy inside the Faraday cage at all. After that, the temperature in the chamber was reduced to absolute zero, –273 degrees C. At this temperature, all matter should (according to quantum theory) stop vibrating and produce no heat. Even the electrons should stop orbiting their atomic nuclei. But what they found instead was there was indeed heat energy still being produced inside the chambers. In fact, there was theoretically enough zero-point energy inside a common electric light bulb to boil all the world's oceans. Because the temperature had been reduced to absolute zero, this energy was dubbed "zero-point energy." These experiments have subsequently been widely discussed as "proof" that there is an as-yet-undetectable three-dimensional energy field that surrounds all things and can be tapped into by some exotic quantum devices. In truth, what these tests actually confirm is the hyper-dimensional model of reality.

What isn't generally understood about the zero-point effect is that even though all motion is supposedly stopped and all electromagnetic interference is blocked inside the cage, the cage itself is far from motionless. It is still sitting on a planet that is rotating once every 24 hours (at about a thousand miles an hour), orbiting the Sun once every 365 days (66,000 miles an hour), and orbiting the galactic center once every 250–260 million years (483,000 miles per hour). So the Faraday cage and everything in it is moving at an extremely rapid clip, in three different motions, which is more than enough to generate Sciama's dynamic torsion field. So the energy inside the cage isn't coming from nothing, from the vacuum; it's coming from the dynamic torsion created by the motion of the Earth itself. And though it's true that no electromagnetic energy can penetrate the Faraday cage from the *outside*, the energy isn't coming from three-dimensional space; it's coming from outside our 3D Universe.

The implications of this are somewhat staggering. Rather than wasting time trying to access "the power of the vacuum," pretty much all you have to do is put up an electrical cell in orbit, and it should generate power from *the motion of the Earth through space itself*.

NASA got a firsthand experience with this in 1996. On space shuttle mission STS-75 a tethered satellite was brought along to conduct an experiment using the satellite and the shuttle itself as two ends of an electric circuit. The idea was to see if electric power could be generated by connecting the Earth's ionosphere with its magnetic field. NASA hoped to someday power the shuttle using such technology, and to see if it could be used as a propulsion system. The satellite, named TSS-1R, was slowly deployed to a distance of about 12 miles from the shuttle. All along the way, the tether was experiencing voltages much greater than NASA's models had predicted. As the shuttle came into the light of sunrise, essentially an alignment with the source of all energy in the solar system (the Sun), an arc of electricity of more than 3,500 volts shot through the system and melted the tether, causing it to snap and the satellite to break away.[1] Unbelievably, even after the tether had snapped and the circuit was theoretically broken, the still-functioning satellite continued to record high levels of electrical energy when there should have been none at all. NASA scientists were expectedly confused by the results. According to the NASA mission scientist, Dr. Noble Stone: *"Things in the data really pop out at you that are [both] unexpected and unexplained.... Electrons detected by the satellite, when current flowed in the tether, were much more energetic than expected."*[2]

Truthfully though, if the guys at NASA had been paying attention to more than 50 years of unrecognized experimental research, they would have had no reason to be surprised by the tethered satellite experiment.

12 The Explorer Effect

So, to the four accepted forces in physics—the strong and weak nuclear forces, gravity, and electromagnetism—I would add torsion. In fact, if it were up to me, I would argue torsion is by far the most important force in nature, and, because we haven't really studied it, the most mysterious. But there have been a number of researchers besides Sciama and DePalma who have been studying torsion for more than five decades now, even if they didn't realize it.

On a windswept Florida beach more than half a century ago, the first successful test of a mechanical torsion field device was conducted completely by accident. On January 31, 1958, the United States finally orbited its first artificial satellite, *Explorer I*. I say "finally," because the simple act of launching a satellite, a process we take almost for granted today, had been an impossible dream for America up until that moment. Engaged in a political battle with the Soviet Union for dominance in space, the "final frontier" as *Star Trek* called it, the U.S. at that time was far behind the Russians. In November 1957 the Soviets had launched *Sputnik I*, the first man-made artificial satellite. This achievement electrified the American public, which had assumed they were well ahead in the "space race." Because of the possible threat of the communists raining nuclear bombs from space with no warning, the United States responded by hurriedly trying to launch a response mission, *Vanguard 1*. It blew up on the launching pad. Desperate, the U.S. turned to Dr. Werner von Braun and his handpicked team of Nazi rocket scientists to find a way to match the Soviets. Von Braun was considered a last resort because of his Nazi past, but he was a genius in rocketry, and he saved the day by successfully putting *Explorer I*, a 30-pound payload, into orbit.

But it almost didn't happen. Or at least, it sure looked that way for a while.

Because of the extremely primitive state of the satellite tracking network at that time, there were only three stations around the world that could detect the signals coming from the orbiting *Explorer I*. Some 90 minutes after launch, it was supposed to be heard by a tracking station near San Diego, confirming that *Explorer I* had attained orbit. When it didn't show up on time, at 12:30 a.m. on January 31, 1958, everyone assumed the satellite had been lost. When it still wasn't heard from 11 minutes later, it was all but certain that the satellite had failed to make orbit. But then at 12:42 a.m., the expected signal came through, true and strong. The delay had been caused by the fact the satellite had somehow, some way, ended up in a dramatically higher orbit than had been predicted.[1]

I cannot overemphasize how impossible this is. There is something called the Rocket Equation that is used to calculate the parameters of a rocket's launch cycle and its predicted orbital arc. I won't bore you with the details, but the fact is a miscalculation of that type simply cannot happen. The thrust of the rocket, the air resistance, and the strength of gravity at various altitudes are all known quantities that are easily accounted for. Yet somehow, *Explorer I* ended up in an orbit that was almost 60 percent higher than it should have been. Despite various conventional explanations being bandied about over the decades since then, none of them have stood up to scrutiny. (See "Von Braun's 50-year-old NASA secret," by Richard C. Hoagland)[2]. However, there is a perfectly viable (though scientifically unacceptable) reason that *Explorer I* ended up so much higher than it should have. It was a torsion effect.

There are two kinds of rocket engines: liquid fueled and solid fueled. Liquids are generally preferred because, by adjusting flow rates of fuel and air with a mechanical valve, you can vary the thrust of the engine to save fuel. Solid rocket engines (or "motors," as they were known at that time) cannot be adjusted because, once they start burning, they just keep burning until their fuel is exhausted. The Jupiter rocket that launched *Explorer I* was designed to have one liquid fueled stages and three solid fueled stages. Because the solid fuel rockets of the era had very inconsistent burn rates, the entire third stage assembly was rotated at 750 RPM. This was done because spinning the whole assembly was a simple way of evening out the inconsistent thrust. Instead, for reasons that must have been a complete mystery to von Braun, it added an enormous amount of energy to the rocket system and *Explorer I* ended up in a much higher orbit (600 miles higher) than it should have. Unable to come to grips with the phenomenon, von Braun and the Americans continued to flail away at the problem for several years afterward. Subsequent missions, like other *Explorer* and the new *Pioneer* missions, exhibited the same weird over-performance. Even the Soviets, who were years ahead of the Americans in guidance and navigation, seemed incapable of executing even the most basic Newtonian maneuvers in space. Their first attempt to "land" an artificial satellite on the Moon—by basically slamming right into it—resulted in a clean miss by almost 3,700 miles—more than the 2,160-mile diameter of the Moon itself. The whole escalating sequence of events came to a head in early 1959 when a JPL-constructed satellite, *Pioneer 4*, missed the Moon by a whopping 37,000 miles, more than *17 times* its diameter.

Now again, this is impossible. Once you escape the gravity of the Earth, it is a very simple calculation to plug in the orbit of the Moon, point and shoot. This is called a "ballistic trajectory," meaning that it's pretty much a straight shot, like a bullet out of a gun. Because there is no air resistance in space and the effects of gravity are tiny in orbit, it should be a piece of cake to hit the Moon from Earth's orbit.

There were even missions that had mid-course corrections built-in to their flight plans that still missed the Moon by more than the proverbial mile. Had this trend continued, the entire space age would probably never have happened, because neither the United States nor the Soviets could seem to find a viable solution.

But Werner von Braun and his team must have eventually realized that if you wanted a spacecraft to follow conventional Newtonian celestial mechanics, Rule One had to be: Don't let it rotate. We can infer this because almost as soon as *Pioneer 4* missed the Moon by more than 17 times its diameter, they immediately abandoned solid fuel rockets, spinning upper stages, and all the related complications of them for the far riskier, and to that point far less reliable, liquid fueled designs.

Choosing to push the rocket design envelope and embark on what was then a radically new type of spacecraft construction, von Braun (in partnership with William Pickering of NASA's Jet Propulsion Laboratory) initiated the *Ranger* series of unmanned lunar probes. *Ranger* was designed to do what *Pioneer 4* had failed to do: aim at and hit the Moon. The *Ranger* missions were to be launched toward the Moon, undergo one mid-course correction, and impact the lunar surface. Running from 1961 to 1965, the wildly expensive (for its time) program was plagued with problems from the outset.

The first two *Rangers*, launched atop an Atlas/Agena liquid fueled two-stage booster rocket, failed to achieve orbit. The third, *Ranger 3*, made it into space but still missed the Moon completely, but this time by a slightly improved 22,000 miles. *Ranger 4*, despite experiencing "an apparent failure of a timer in the spacecraft's central computer," somehow managed to impact the lunar surface on the far side. It must have seemed von Braun was getting closer to solving the riddle.

But then *Ranger 5*, which should have been a bull's-eye, skipped past the Moon at a distance of just 450 miles. *Ranger 6* managed to impact the Moon precisely where it was supposed to, but its

TV cameras failed. It was not until *Ranger 7* in 1964 that a *Ranger* mission could be considered to be completely successful, having achieved all the stated mission objectives and hitting the Moon exactly where it was supposed to.

So what had happened? Had von Braun somehow managed to solve the *Explorer* problem by trial and error? And if the simple act of abandoning the rotating upper stages was really the key to the riddle of the Explorer Effect, why did *Ranger 3* still miss the Moon by 22,000 miles? It didn't have the rotating upper stages like the *Explorer* missions' Jupiter rockets had. As you'll see, *Ranger 3* is actually the exception that proves the rule.

As Richard Hoagland has written on his Website, Werner von Braun had been adamantly opposed to the concept of Lunar Orbit Rendezvous (LOR), then a popular proposal for how to get men to the Moon and back safely. LOR involved launching a rocket into Earth orbit, sending two separate vehicles (an orbiter and a lunar landing vehicle) to the Moon, where they would detach, and the lander would then land and then later rendezvous in lunar orbit for the trip home. Von Braun instead advocated a far more costly and difficult method called "Direct Ascent." Direct ascent meant that the rocket was launched, went to the Moon, landed, and came back all in one piece. This stubborn insistence on an infinitely more expensive and risky strategy for getting to the Moon made no sense except in the context of the *Explorer* anomaly. If the Effect was that dramatic just getting a small satellite into Earth orbit, the notion of rendezvousing in Earth orbit, traveling to the Moon, and then rendezvousing again was all but unthinkable. But then suddenly, to the shock of all those around him, von Braun reversed his position and embraced Lunar Orbit Rendezvous at a design review in the summer of 1962. The reason, it is now obvious, was *Ranger 4*.

Launched in April 1962, just a few months before the critical LOR design review, *Ranger 4* managed to do what no other American spacecraft had done to that point: aim for and impact the Moon. This must have given von Braun the confidence that he had solved the problem of the *Explorer* over-performance, and could compensate for it. But the reality is that *Ranger 4* shouldn't have hit the Moon at all. By all reasonable measures, it should not have been able to hit the broad side of a barn. You see, *Ranger 4* was designed to be boosted into space on a ballistic trajectory (shot like a bullet in a straight line) to the Moon and then hit it head on. A mid-course correction was built into the mission profile. Interestingly, if there was no true "Explorer Effect," then such a mid-course correction shouldn't have been necessary at all. You just line the rocket up and fire the spacecraft to the target point, and the Moon should be there, right on schedule.

And that's exactly what happened. But it shouldn't have.

The problem is that *Ranger 4* developed a radio malfunction immediately after launch, and was not able to execute its mid-course correction or even deploy its solar panels. As a result, it ran out of power just a few hours after launch. So how then did it impact the Moon, when *Ranger 3*, which had no such malfunctions, missed by some 22,000 miles?

The answer is that they hadn't stopped the rotation on the Rangers at all.

After missing the Moon with *Pioneer 4* by 37,000 miles and by "only" 22,000 miles with *Ranger 3*, von Braun must have felt he was starting to get the hang of it. He must have assumed that, when they stopped the rotation of the spacecraft, it would be a relative piece of cake to hit the Moon, because all of the anomalous over-performance would have been squeezed out of the system. It must have come as quite a shock when it turned out that *Ranger 3* had, just like the *Explorer* and *Vanguard* series' before it, somehow developed "a malfunction in the booster guidance system [that] resulted in excessive spacecraft speed." Despite a mid-course correction maneuver, the spacecraft failed to get any closer to its target. In fact, according to later analysis, the maneuver did nothing to alter the spacecraft's trajectory at all. It had started out some 22,000 miles off course due to the "excessive spacecraft speed," and it remained some 22,000 miles off course despite the mid-course correction. (Again, this is impossible according to the accepted laws of physics.) So, where had *Ranger 3*'s excess boost come from, if the booster wasn't rotating, and why hadn't *Ranger 4* behaved the same way? Why had it managed to hit its target virtually dead on?

The reason is because even non-rotating spacecraft and rockets, in their boosters and in their instrument packages, always contain at least one thing which is always rotating, even if the vehicles do not: the gyroscopes.

All spacecraft and their associated launch vehicles have a number of whirling gyroscopes in their on-board inertial navigation systems. These devices, along with other on-board devices, called accelerometers, literally steer the vehicles, providing on-board 3D coordinates for ground-based navigation. They provide absolutely critical reference points for any spacecraft trying to reach any distant destination. But in order to do this, the gyros always have to spin. And, they spin much faster (10,000 RPM at least) than the sluggish 750 RPM of the Jupiter rocket's upper stage. Experiments by DePalma in the 1970s confirmed that a smaller system like the gyroscopes, which are spinning faster, actually generates a much more dramatic torsion effect than a larger system that is spinning more slowly.

So when *Ranger 3* missed the Moon by 22,000 miles (in spite of the ill-fated mid-course correction) it was operating normally, it was powered up, and its gyros were rotating. When *Ranger 4* hit the Moon almost dead on—from 240,000 miles, with no mid-course correction—it was a dead spacecraft, it was drained of power, and its gyros were flat-lined. It was little more than a bullet on a ballistic trajectory, with no troublesome spinning gyros to add energy to the system and deflect the course it was on.

It had to be at that moment in April 1962 that Werner von Braun realized that the Explorer Effect would apply even if the spacecraft itself wasn't spinning, but only the gyros were. Furthermore, these two missions, using virtually identical hardware, would have given him the crucial data he would need to calibrate exactly what the corrections had to be in his guidance equations. He had everything he needed to compensate for the Explorer Effect and get men to the Moon and back. From that point on, he could embrace the far cheaper Lunar Orbit Rendezvous scenario, because he now knew what he had to do to account for the Explorer Effect. And for that piece of test data alone, the wildly expensive *Ranger* series was worth its weight in gold.

But even though he had solved the riddle of spacecraft guidance, von Braun clearly had suspicions that something was very wrong with the basic laws of physics. We know this because almost immediately after *Explorer I*, he began to seek out some very interesting counsel.

The Allais Effect

Maurice Allais is a Nobel Prize–winning French economist who won the award "for his pioneering contributions to the theory of markets and efficient utilization of resources" in 1988. Besides his ongoing interest in economics, Allais also at various times during his storied academic career turned his attention to dabbling in physics. In the late 1950s, he made a series of observations during a total solar eclipse in France that shook the scientific world to its core.

Allais's original interest during these experiments was to test some of Einstein's ideas about the linking of gravity and magnetism. (In other words, he was conducting some experiments into the unified field theory.) "My main idea at the start was that a link could be established between magnetism and gravitation by observing the movements of a pendulum consisting of a glass ball oscillating in a magnetic field," Allais said in his Nobel Prize acceptance speech. What he discovered astounded him and electrified the French science community.

Over the course of more than a month, Allais and a team of dedicated researchers took turns releasing a paraconical pendulum, a free swinging device that he invented that (obviously) moves back and forth in two perpendicular directions. He got the idea from what is called a Foucault pendulum, a similar device that was invented by French physicist Léon Foucault (1819–1868). A Foucault pendulum was first used to demonstrate the rotation of the Earth in 1851. The principal is that the pendulum is set to swing back and forth in a fixed plane, while the Earth rotates underneath it. Allais's paraconical pendulum was different in that it freely swung *with* the rotation of the Earth. To his (and everyone else's) surprise, during the 1954 partial solar eclipse over Paris, the pendulum, rather than continuing to swing with the rotation of the earth as Foucault's had, suddenly started to rotate very rapidly *backward, against* the rotation of the Earth, and continued to do so for the duration of the eclipse. Once the eclipse ended, the pendulum resumed its original motion in resonance with the rotation of the Earth.

In a closed, Newtonian/Einsteininan model of reality, this is simply impossible. There is no known force that can operate on a pendulum from a distance in this manner: not gravity, not magnetism, not electrical forces. Needless to say, this result sent shockwaves through the fairly small French science community that was aware of it. Several possibilities to explain why the experiment didn't violate the known laws of physics were suggested, including that high altitude winds had somehow effected gravity during the eclipse or that air pressure changes because of it had effected the experiment. In response to this, Allais (who was quite confident of the validity of his results) set up another identical experiment for a 1959 partial solar eclipse, but this time, he did it a specially constructed environmentally controlled lab almost 200 feet underground. The results were the same.

It's hard to overstate how significant Allais's findings really were. He discovered that, not only did the alignment of three celestial objects (the Sun, the Earth, and the Moon) cause his pendulum to behave in a bizarre manner, it did so without any possible explanation as to what force could have been acting on it. Of course, we can now conclude that it was a torsion effect, caused by the pendulum's rotation (the back and forth movement is partial rotation) and the rotation of the Earth itself. His experiment is also more proof of the existence of the Aether, because not only is there is no scientifically accepted force that could have affected the pendulum's motion from a distance, there is also no known medium through which this force could have propagated other than the Aether. Allais himself has called for all of Einstein's work on relativity to be re-examined, because his experiments completely undercut the "laws" of gravity.

So like Nelson's RCA study before him, Allais's experiment showed that astrology has a basis in physical reality (the eclipse alignment is similar to a conjunction in astrology). He also, in further experiments, discovered something even more interesting: The effect was not isotropic—it did not propagate equally in all directions. It was what scientists call anisotropic, meaning that it traveled in a very specific direction, from the Sun to the Moon to the Earth.

Through the years, there have been a few attempts to duplicate Allais's findings, with mixed results. In 1999, NASA physicist Dr. David Noever planned to set up experiments to test the effect during an eclipse in August of that year. He succeeded in getting several European laboratories (and one in China, which was in the path of the eclipse) to set up pendulums. Allais wrote a long briefing paper for Noever, in which he explained that the experiment had to be done in a very specific way, with very specific equipment, and with very specific protocols. He lamented the results of some of the more recent attempts in the '70s and '80s to confirm his findings, even dancing around the idea that many of them had been set up to fail from the beginning.

"Science has lost at least [40] years. Not only have my experiments not been [properly] followed up, but they have been successfully hidden," he wrote.[3] And it's hard to argue with him given what happened next.

Instead of following Allais's suggested protocols, Dr. Noever ignored him and set up experiments in several places around the world using the type of pendulum that Allais said would most likely *not* produce a positive result. After the eclipse, which NASA had touted on its Website[4] as being sort of the final answer on the so-called "Allais Effect," Noever simply collected the data and dropped off the face of the Earth. He had been scheduled to write a contributing article about the experiments in a compilation book about gravity from Aperion books,[5] but according to the editor, Matthew R. Edwards, he never submitted it. *"Yes, unfortunately the article by David Noever never materialized. I don't know why; he just stopped communicating. Apparently, his European colleagues had the same experience with him. They claim he has left NASA and took all the eclipse data with him! Does anyone know the background story, and where the 1999 Eclipse data is???"*[6]

Noever and members of his team eventually re-emerged at a private Internet company, and he has made almost no public comment about the results of the eclipse experiments of 1999. Some results were eventually published in their native countries, and, despite using the wrong type of pendulums, two experiments reported positive results while one concluded that "Observed deviations are within the expected tolerances and cannot be seen as anomalies."[7] That statement I find kind of curious. See, the problem is that there shouldn't be *any* "observed deviations," at all. Pendulums *can't* rotate backward. Period. Translation: "Oh yeah, we got the same effect, but we are downplaying it."

The interesting thing to me is that almost no one in the English-speaking world would have a clue who the hell Maurice Allais is or what his experiments showed if weren't for one very diligent individual: NASA's Dr. Werner von Braun. Shortly after the bizarre extra energy effect of the *Explorer I* launch, von Braun began to quietly inquire all over the world about any experiments involving gravity. (Note: The Allais Effect, like the Explorer Effect, has nothing to do with gravity per se, but at that time, because no one really had a clue about torsion or hyper-dimensional physics, the only force that anybody could think of that could cause the observed phenomena was gravity, so that's where von Braun went looking). He first contacted West German physics theorist Burkhard Heim, who had written a paper on using certain unified field equations to create a "fuelless field propulsion system" for space travel. Von Braun quickly found out that Heim had no money to conduct experiments, so he then turned to Allais, and arranged to have his French research papers published in the United States in English. It's quite clear from these actions that von Braun knew there was something radically wrong with the current models of physics, and it's equally clear he worked out a way to get his rockets to account for the extra energy his spinning systems were creating, because we did manage to get to the Moon and back, and we can today rendezvous and dock two spacecraft with relative ease. But that was hardly the case at the dawn of space age.

So it takes a fairly simple process of detective work and deduction to figure out how the United States managed to conquer the Explorer Effect and get men to the Moon and back. What was, until recently, a mystery is exactly how the Russians figured out the same thing.

Nikolai Aleksandrovich Kozyrev

Nikolai Aleksandrovich Kozyrev (1908–1983) was a Russian physicist of some considerable renown in the early part of the 20th century, the "golden" era of physics. His first scientific paper was published when he was only 17 years old, and by his late 20s he was one of the leading Russian thinkers of his day. In 1931, he began work at the Pulkovo Observatory south of what was then Leningrad (now St. Petersburg) Russia. The Soviet Union by that time was being ruled by the tyrant Joseph Stalin, one of the most vile men to ever walk the face of the Earth. In 1936, a disgruntled graduate student accused certain members of the faculty at the observatory of being disloyal to the State and Stalin, and the entire faculty, including Kozyrev, were arrested and thrown into one of the horrific Soviet gulags. He spent 10 tortuous years in the prison camp, suffering all manner of deprivation and abuse, and was the only one of the scientists arrested at Pulkovo to survive life in the Soviet concentration camps.

Kozyrev, like many victims who suffer some form of abuse, be it physical, mental or sexual, withdrew from the material world and began to leave his body, in essence spending much of his time in prison in the deep isolation of his own mind.

There are two life-changing episodes that took place during Kozyrev's confinement that testify to the extent of his suffering and the spiritual state of mind it induced. While being held in a cell in the central prison in the town of Dmitrov, Kozyrev had been musing on the source of the inner heat of the Sun. After a terrible experience that resulted in the death of his cell mate, Kozyrev went back into his emotional shell and realized that in order to solve the problem he was contemplating he needed some data from a book that was part of the library at the Pulkovo observatory. *The Course of Astrophysics and Stellar Astronomy* was hardly the kind of reading material available to prisoners in the camp, but one day, without warning, an unseen hand slid the book through the service slot in his cell. He was able to keep the book for several days before the guards discovered it, and was also able to commit the necessary passages to memory before it was seized. Consumed with his meditations on the subject, Kozyrev then began to unconsciously pace in his cell, an activity that was forbidden by the cruel guards. (Prisoners were required to be seated on a stool at all times during the day.) Infuriated at what they saw as a second act of defiance, the guards placed him in solitary confinement for five days, a sentence that had just resulted in the death of his cell mate a few days before. Stuck in the tiny, dark prison cell in the dead of the Russian winter, barefoot and clothed in only his underwear, Kozyrev tried desperately to survive in the zero-degree temperatures while eating only a single piece of bread and getting one cup of hot water per day. Finally in his despair he began to pray to God, and as he described it later an inner light began to heat his body. Without this miraculous external source of heat energy warming his body, Kozyrev would surely have not survived his ordeal. When he was eventually released from solitary confinement he set about the task of trying to understand the true nature of God's creation, and of the source of the mysterious heat energy that had saved his life.

Further escaping from the horrors around him and the deprivation of his imprisonment, he began to muse on the true nature of the world and the Universe itself. One of the things he considered were the many patterns of life, most especially the beauty he saw in nature. As he contemplated, he started to see that physical life had a distinctive pattern, the spiral, and he began to understand that it must be driven by some underlying force of nature. He saw that many forms of life, like the nautilus shell and even simple protoplasm's, grew in this same spiraling pattern. What he decided was that this spiraling motion, which is obviously torsion at work, was nothing less than *the physical manifestation of time*. He realized that time was far more than just a measure of duration. It was a physical, forward movement, without which nothing could exist. He began to understand that Einstein's "space-time," the fabric of the Universe itself, twisted and spun and spiraled in constant motion, and that it was this constant forward growth that made reality a thing at all. Everything, as he saw it, came from motion. Without motion, there is no energy, and without energy there is nothing. No history. No future. No movement at all. So a complete lack of motion—of forward spiraling movement—meant that time itself would stop.

As abstract as this idea sounds, consider this example. If a hurricane blows down a palm tree next year in Florida, that is the result of movement. The movement of the Earth through the heavens creates the energy that feeds the winds. The winds feed the hurricane, and the palm tree being blown over is the future event, the time element, which is the result of all that constant forward motion.

He eventually came to see that this spiral motion was the natural growth pattern of all living things, and he resolved to spend the rest of his days studying the fundamental insight he had been granted during his suffering in the prison camps. Once he was finally released from prison in 1946 and considered "rehabilitated," Kozyrev turned his attention to finishing a doctoral thesis begun before his imprisonment.

The subject he chose was the source of the same inner light that had saved his life in the Dmitrov prison camp. Because such a personal experience could hardly be used as the subject of a doctorate of physics dissertation, especially in the repressive Soviet Union, Kozyrev elected to address the source of the inner heat of the Sun. In the intervening years since his imprisonment, the idea that the Sun was fueled by constant ongoing nuclear fusion had taken hold. Undaunted by these proposals, Kozyrev argued against the stellar fusion theory on several counts. First, if thermonuclear reactions were indeed taking place in the core of stars like the Sun, then the internal temperature of the Sun should be about 20 million degrees. Kozyrev's calculations showed it to be at best only about 6 million degrees, the same as the temperature at its surface. Second, he pointed out that if the entire Sun was thermonuclear furnace, then the amount of energy it put out—its surface temperature—should be far greater than 6 million degrees given its mass. Kozyrev's conclusion from his work was that stars like the Sun are not nuclear fusion furnaces at all; they are spinning energy portals, energy machines, and the heat they put out is simply a function of regulating the internal heat to keep it from over producing energy.

Through the years, Kozyrev has been both lauded and criticized for his theories, but to this day his work on the source of the Sun's inner light has yet to be disproven. One of the keys to validating the idea that stars are just nuclear fusion furnaces is the presence of particles called neutrinos. If nuclear fusion is indeed taking place in the stellar core, than the Sun should be pumping out a predictable flow of neutrinos as a byproduct of that reaction. When the first measurements of the Sun's neutrino emissions were made, it was found to be putting out only about one third of the expected neutrinos. Later observations showed no change. So again, as they had done with dark matter and other scientific problems, instead of accepting that there was something fundamentally wrong with the standard solar model, the astrophysicists instead set about "proving" that neutrinos change from one type when they are created (the type that should be detected) to another type. By this tortured logic, modern physicists have now concluded if you add up all the neutrinos emitted from the Sun in all the categories—even the ones that don't come from stellar fusion—it still proves that all of the Sun's energy output comes from thermonuclear fusion. How exactly these little neutrinos magically change their composition from one type to the next is never really explained in any of these papers, but, again, when you are dealing with a threat to the established orthodoxy, this is the kind of logic you can expect.

Kozyrev did devote himself to this central question for years after his release, conducting experiments that showed that the source of the Sun's heat energy was function of *time*. Using his concept of time, what he was really saying was that the Sun's energy came from its spin and its forward motion; in other words, in Kozyrev's model, time and torsion are the same thing. They are a function of motion.

Sadly, what he did not live to see, even after all that he went through, were a series of experiments that proved that his ideas, along with Allais's and DePalma's and von Braun's, were not only scientifically valid, they were utterly relevant to our current era in the flow of time.

13 The 2012 Venus Transit

As we learned in Chapter 8, the Mayans were fascinated with Venus and, more specifically, with its infrequent transits across the disk of the Sun. In astronomical jargon, a "transit" is essentially a mini-eclipse in which a planet or other body passes between the Earth and the Sun. Transits of Venus are very rare. In fact, before 2004 there had been only six Venus transit cycles since the invention of the telescope: 1631, 1639, 1761, 1769, 1874, and 1882. Because of the oddities of Venus orbit, these transits occur only every 243 years, with pairs of transits eight years apart separated by long gaps of 121.5 years and 105.5 years. In other words, Venus passes in front of the Sun on two occasions, eight years apart, separated by 121.5 years and then 105.5 years. Or, there are four Venus transits every 243 years. The last such transit was a spectacular display recorded on June 8, 2004, by telescopes and cameras all over the world. The next transit of Venus occurs on June 5–6, 2012, just a few days after a solar eclipse and obviously the same year that the Mayan Long Count ends. Now, given the Mayan proclivity for tracking the motions of Venus in the Dresden Codex and the fact that the codex refers to the start of the current Long Count cycle as the "birth of Venus," it seems just a little improbable to me that the Long Count calendar cycle ends in the same year as such a rare astronomical event. Just as John Major Jenkins and others argue that the Mayans must have known that the Long Count ended on the 2012 winter solstice, so must they have known it ended in the same year as a Venus transit. They were just too smart to have missed it.

Clearly, the Maya considered the Venus transits to be important, but the obvious question is why. In looking back over the many historical Venus transits, they always seem to occur in periods of great change and social upheaval. A pair of Venus transits occurred in 1518 and 1526, the very period where Hernan Cortez landed in Mexico and conquered the ancient Aztecs. Because of the circumstances, the Aztecs had been expecting the feathered serpent god Kukulkan to return and basically laid down for Cortez, thinking that's who he was. The next transits were in 1631 and 1639, and were followed by a near complete suspension of the sunspot cycle. This lasted for about 70 years and coincided with the "little ice age" that occurred between 1645 and 1720. The next transit pair occurred in 1761 and 1769, which coincided with the early events of the American Revolution. The period between the 1874 and 1882 transits saw the rise of the early New Age movements and fevered interest in all sorts of occult societies, including the Theosophical Society and Order of the Temple of the Rosy Cross. What this implies is that just as I alluded to in Chapter 2, there might actually be an underlying physical affect here on Earth that is driving these behavior cycles.

If we look at the 1631 and 1639 transits, it was after that cycle that the Sun seemed to go dormant. In the transits in the late 1700s and in the 1874–1882 transit periods, we see instead a change in human thought processes. Perhaps there is a pattern of conscious change in between the transits and physical change *after* the transits. If that is the case, then we should be in the midst of such an energetic shift right now.

According to some Mayan scholars, Venus represents the so-called "divine feminine," an astrological concept that argues that the female energy of love, nurturing, and compassion will replace the more male characteristics associated with conflict and national identity. They see the Venus transit of 2012 as a marker that closes the loop between the birth of Venus, the beginning of the current Long Count cycle, and coming of the new female energy of the New Age. Venus has always been associated with such Goddess energy as well as love, beauty, art, and the heart. I do find it interesting that there seems to be a shift in our thinking in the years since the 2004 transit. More and more Americans have tired of the overseas wars, and the economic instability is being driven by an underlying lack of faith in our current monetary system. Suspicion of the motives and honesty of those in government is at an all-time high. Certainly, like the 1761–1769 Venus transit cycle that pre-dated the American Revolution, change seems to be in the air. A growing number of Americans are questioning the very basic tenants of our society in a way that they have not done since at least the Great Depression, if not the Revolution itself. Does this mean that Calleman, Jenkins, and the other Mayan scholars who argue that the Codex and the calendar are just a road map of human consciousness are correct? We'll get to that question later, but more worrisome to me is the possibility that the Venus transits of 1631 and 1639 somehow affected the Sun's energy output.

If, as the argument goes, the end of the calendar cycle is somehow linked to solar output, what would be the physical mechanism? Again, could it have something to do with the transit of Venus as opposed to the influence of Jupiter and Saturn, as some scientific papers have argued? We already know from the Allais Effect that eclipses cause subtle vibrations through the hyper-dimensional Aether, which can be measured by our instruments, like Allais's pendulums. With this in mind, my co-author on *Dark Mission*, Richard C. Hoagland, conducted a series of experiments in 2004 and in 2006. What he hoped to do was establish once and for all that planetary alignments, like Allais's solar eclipses over France and the coming

2012 Venus transit, can and do create a "tremor in the force"—a measureable physical torsion effect that can change our thinking, and perhaps our planet.

To do this, he started by reviewing experiments done by Dr. Bruce DePalma using tuning forks. DePalma had discovered that tuning forks were very sensitive to the waves of torsion energy that were generated by his rotating test rigs. This is because vibration, a rapid motion back and forth, is really just partial rotation, and, as we've established, rotation is the key to releasing this pent-up hyper-dimensional energy. DePalma used tuning forks that he got from a men's wrist watch of the period called the Accutron because it was one of the most accurate and precise timekeeping devices of the era, losing only about one minute of time per month.

The site chosen for the 2004 experiment was southeast Florida's famed Coral Castle (an entire story in itself). This location was selected in part because toward the very end of the 2004 Venus transit, just as Venus was about to pass beyond the disc of the Sun, the Sun itself would be rising over the horizon of the Atlantic Ocean. Because this would re-create an alignment similar to the NASA tethered satellite experiment, the physical sunrise seemed like an excellent time to be measuring the possible torsion effects. What he found was that the measured effect was not only powerfully obvious, it was astonishing. As Venus crossed the solar disk, the tuning fork vibrated much faster—because it was receiving more energy through the Aether from the effects of the transit. But what is most interesting as you look at the data is just where the maximum energy spikes take place. They are on what are called the "contact" conditions.

In geometry, two objects are "tangent" to each other when one of each of their edges is lined up, or touching. These same geometric conditions are called "contacts" in astronomy, and it basically means that one edge of Venus is "touching" one edge of the Sun, at least from our vantage point here on Earth. When the outside edge of Venus is touching the outside edge of the Sun, it's is called an "external contact." When the inside edge of Venus is tangent to the edge of the Sun, it is called an "internal contact." In the case of the 2004 transit, the effects on the tuning fork were most dramatic as the edge of Venus's disk passed in front of and aligned with the visible edges of the Sun. The experiments showed that these specific edge alignments caused the most dramatic disruption of the normal behavior of the tuning fork.

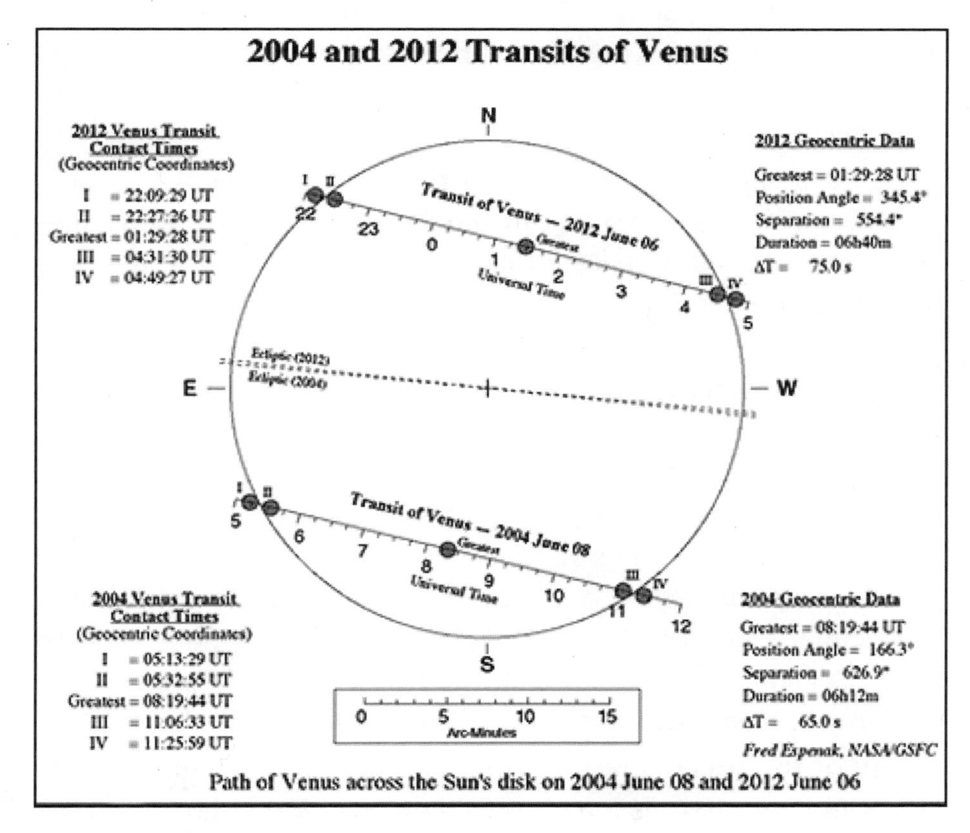

Image 13.1. The Venus transits of 2004 and 2012, with contact points illustrated. Image courtesy of NASA.

When Venus first touched the edge of the Sun, creating its own mini-eclipse, the little tuning fork went wild. As predicted, the contact conditions and the alignment of Venus's spin axis with the limb of the solar disk generated the strongest responses, well beyond the margin of error. In fact, the most significant frequency spike of the Accutron tuning fork occurred exactly at the moment of third contact, when Venus's edge just "kissed" the Sun's eastern limb. It went from a relatively quiet 360 hertz (cycles per second) to 364.5 hertz. If this tuning fork had been in an actual watch, the acceleration would have worked out to about 12 minutes per day—for a watch normally accurate to within one minute per month. In other words, just like Allais's pendulum effect, something had reached out from Venus at the precise moment of the contact and "touched" the little Accutron detector from 25 million miles away.

It was a vivid confirmation that planetary alignments of major members of the solar system can and do have extremely powerful—and physically measurable—effects on Earth. And, these waves of torsion energy can be dramatically increased in intensity if certain geometric configurations are in place.

In all the measurements taken during the transit, the most powerful effects were at the third (internal) and fourth (external) contacts, and they continued resonating for quite some time after Venus had cleared the Sun. What this means is that the physics is most affected—the tremor in the force is greatest—as the alignment comes to an end, and it reverberates for quite a while thereafter. This reminds me a great deal of the Princeton Global Consciousness Project's measurements during the 9/11 attacks—the strongest energies affected the instruments *after* the attacks had ended.

All of this was interesting, but perhaps not compelling. Truthfully, although there is a measureable effect, and although it does defy all the established laws of physics, it wasn't really powerful enough to cause a significant change in our physical environment or our emotional experience, at least not consciously. The changes in the vibration of a very sensitive little tuning fork are quite a different matter than shifting magnetic poles, toppling the Earth on its axis or even causing a second American Revolution. But, we should remember that every cell and every atom in our bodies is vibrating, and when we receive signals from on-high it is most likely that they will change our consciousness in ways that are very subtle, just like these. However, the next set of experiments had a great deal more relevance to the questions at hand, and the 2012 transitional period.

After the success of the Coral Castle experiments, Richard decided to follow these up with another set of experiments based out of his hometown of Albuquerque, New Mexico. Setting up his tuning fork test rig again, he decided to test if the same effect could be detected around a much bigger and more relevant stellar alignment, the alignment of the Sun, the Earth, and the galactic core.

Galactic Alignment 2012

As we've now established, celestial alignments such as solar eclipses and Venus transits can and do have a measurable effect on sensitive instruments here on Earth, a scientific impossibility according to mainstream physics. Neither gravity, magnetism, nor any other accepted force of nature could create these effects, but they are demonstrably real. Only torsion, of which we know very little experimentally simply because it hasn't been studied in any significant way, could possibly influence the instruments during these alignment events. Theoretically, the "galactic alignment" of December 21, 2012, would dwarf the fairly miniscule effect observed in the Coral Castle experiments.

The Mayan calendar Galactic Alignment theory was first proposed by John Major Jenkins in his book *Maya Cosmogenesis 2012*. Because of the cycle of the precession of the equinoxes, the Sun is seen to rise in a different house of the zodiac on the winter solstice every 2,160 years. Currently, the Sun rises in the constellation of Sagittarius, which as we have already learned houses the star Sagittarius A and the supermassive black hole at the center of the galaxy.

In *Maya Cosmogenesis 2012*, Jenkins argued that end of the Long Count calendar cycle was tied to a very rare and specific astrological event: the alignment of the Sun with the galactic equator and the galactic center on the winter solstice of 2012. This alignment has not occurred for almost 26,000 years (25,920 to be exact), making it not only an extremely rare event but also one that can be used to mark the end of a single precessional cycle and the beginning of another Great Astronomical Year. Using his own interpretation for the center axis of the Milky Way galaxy, Jenkins concluded that the Sun would come into alignment with that axis and the galactic center on the winter solstice for a period of 36 years, from 1980 to 2016. That the Mayan calendar Long Count ended in 2012 on the winter solstice near the end of this period seemed to him too much of a coincidence, and he concluded that the Long Count coinciding with this astronomical alignment was intentional.

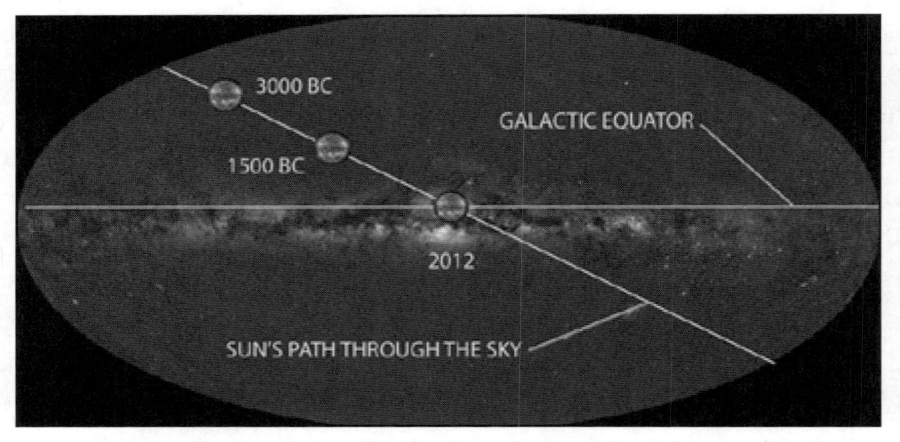

Image 13.2. The Sun being "birthed" in the mouth of the great dark serpent at the center of the galaxy on December 21, 2012.

The exact center plane of the Milky Way passes through a region called the Dark Rift by astronomers and the Dark Road by the Maya. This area in the central plane of the Milky Way is made up of massive amounts of dust and gas that obscure much of the starlight from behind, leading to a perception of a dark serpent coiling around the galactic center. At the center of the galaxy near Sagittarius A, this dark rift opens up, creating the illusion of the serpent opening its jaws to consume the center of the galaxy and its own tail. This poetic view of the heavens is the inspiration for the Ouroboros symbol of many esoteric societies. On the morning of December 21, 2012, the Sun will rise literally in the mouth of the Dark Serpent, as if being born again from some immense cosmic womb. Jenkins argues that this symbolic rebirthing of the Sun had enough significance to the Maya that they marked the end of Long Count calendar by it, and by implication they knew it was a cyclical event that repeated roughly once every 26,000 years.

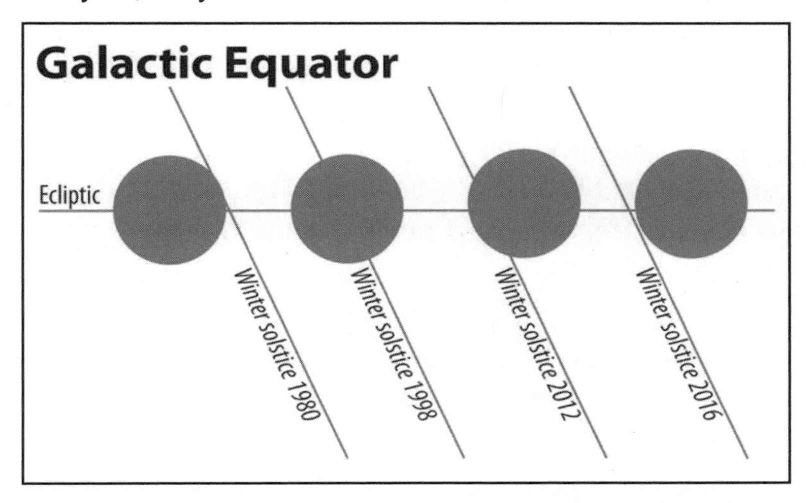

Image 13.3. Alignment of the Sun with the Galactic Equator during the 1980–2016 period. Best alignment with the Equator took place in 1998. Image by Krys Lilly.

Understanding the importance of such alignments in hyper-dimensional physics theory, Hoagland set up an identical test rig to the one he used at Coral Castle in Florida for the Venus transit of 2004. Astrologically, the Sun is in Sagittarius December 18th to January 18 every year. Theoretically, every December–January period leading up to 2012 (and at least to 2016) should create a month long opportunity where the alignment of the Sun and the galactic center should be generating torsion waves that can be detected. As we learned earlier, in hyper-dimensional physics, it is not mass or gravity that count, but spin energy—angular momentum. The black hole at the center of galaxy contains enormous amounts of momentum. Estimated to be 27 million miles across (greater than the distance of Venus at the closest approach) it is also calculated to be 4.3 million times more massive than our own Sun, which is 334,000 times more massive than the Earth. Because gravity is tied to mass the amount of gravity generated by the black hole is beyond comprehension. But, we also know that gravity dissipates at a distance, and, because we are 26,000 light years away from the center of the galaxy, it cannot affect us here on Earth. However, studies have found that this supermassive black hole is also spinning so fast that it rotates once every *11 minutes*. By comparison, the Earth rotates once every 24 hours, and the Sun once every month; and the Sun is less than 1/27th the estimated size of the black hole and 4 million times less massive. So at that size, mass and rate of rotation, how much torsion spin energy is that black hole at the center of the galaxy generating? More than enough to not only hold the entire galaxy together and feed it with life energy, but also certainly more than enough to affect us here on Earth in the manner that the Hindu myths of the Vishnu Nabhi suggested.

Given all the different catastrophe scenarios around the 2012 period, solar flares, magnetic pole shifts, and crustal displacement, Richard knew that his experiments would be important. But all of these scenarios lacked one crucial component: causality. There was nothing in mainstream physics that could make the 2012 solar alignment with the galactic core cause any of them to occur. But if there was an unknown energy pulse coming from the galactic core, and if it could be identified as a torsion effect, then all of the doomsayers might finally have some ammo for their "end is nigh" proclamations. So in December of 2006, he set his equipment up and started taking readings.

The first sample was set up on the winter solstice, December 21st, looking for any anomalous readings. The tuning fork behaved quite normally for most of the day, averaging out at 360 hertz (back and forth cycles per second). Then suddenly, late in the day, there was a completely unexplained spike in the data. For a period of more than 11 seconds (well outside the margin of error) the readings suddenly jumped from 360 cycles per second to more than 3,360 cycles per second, a 900 percent increase in the vibration of the tuning fork. As expected, this sudden surge in energy, which is completely unexplainable by any known force acting on the little tuning fork, dwarfs the subtle readings detected in the Venus transit experiments or Allais's pendulum observations. Although this was exciting, it still wasn't anything more than an academic exercise, because the effect certainly didn't cause any solar flares or earthquakes.

Then he got an inspiration. As a control for his experiment, he continued to take readings for the next month. On several dates, he got similar vibrational spikes in the tuning fork detectors. On New Year's Day, January 1st, there was a 17-second spike of about 219 hertz. Much smaller than the winter solstice reading, but still a 60 percent increase over the normal vibration cycle and it was also of longer duration. Then on January 10th, there was a 40 percent jump in the vibration speed of the tuning fork that lasted for more than *four minutes*. Finally, on January 19th, there was a 42 second spike in the data from 360 hertz to over 7,500 hertz, a 2,100 percent increase in the frequency of the forks' vibrations!

Obviously, these results were off the scale in terms of subtlety of the effect and margin of error. Not only that, but they also showed that in the period after the winter solstice, from December 21st into January, the effect got dramatically stronger. Now, critics of the 2012 alignment scenarios will point out that the best alignment with the Galactic Equator was on the solstice, as it is every year, and not in January, so this effect cannot be connected to a burst of energy from the center of the galaxy. They will also argue that if the effect is real, why does it only last for a few minutes or seconds on some days? Both of these issues are actually fairly easy to address.

If you go back and look at the Venus transit graphics (Image 13.1 on page 163) you'll be reminded that the most dramatic effects on the tuning fork were at the end of the Venus transit and after it had cleared the Sun. Moreover, the effect continued to cause the fork to vibrate and even increase for hours after Venus had cleared the Sun. This implies that a similar effect on a bigger scale (months instead of hours) is taking place with the alignment with the galactic core, and that it has something to do with the geometry of the energy burst itself. DePalma, in his experiments with torsion forces, found that the effect was very specifically directional, as in a beam of energy instead of a wave. Allais found the same thing in his pendulum experiments. So what Hoagland theorized was that the torsion beam coming from the galactic center was also directional, like the light beam from a lighthouse. Only when we are directly in its path does it affect our instruments. Pulsars and quasars, rapidly spinning collapsed stars, emit radio and x-ray energy in the same pattern, in very tight beams of energy that also twist and spin, just as the torsion theory would dictate they do. Using the data points he had collected, Hoagland was able to map out the axis of this energy burst based on where the Sun was in relation to the galactic center on each of the days.

These data points also revealed a timing pattern. About once every nine days, the galactic core, or more likely the black hole at the center of it, sent out a pulse of torsion energy. And each time, it got progressively stronger as our Sun and Earth got more in line with center axis of the beam. This implies that it is January 2013, not necessarily December 21, 2012, that is the time we should be most vigilant and concerned about physical effects on our planet. And, given the fact that 2016 is actually equivalent to the fourth contact event of the Venus transit, we should perhaps be more concerned about 2016 and beyond than we are with 2012.

But again, this scenario is fine as far as it goes. Critics still argue that, if 1998 is the best alignment with the Sun on the galactic equator, then that is when we should have had the strongest physical effects. And they may have a point. After all, December 21, 1998 came and went, and nothing at all unusual happened. Did it?

Well, not exactly. Besides the Earth's large precessional wobble on its axis—the one that takes 26,000 years to complete—there is a second, more subtle wobble called the "Chandler wobble." Officially, this small wobble is attributed to changes in the ocean pressures at great depths as well as shifts in atmospheric pressures. The Chandler wobble tends to vary somewhat, reaching a peak in 1910 and completely stopping for a few months in 2006. However, in December 1998, just as the Sun and our solar system were coming into their most perfect alignment with the galactic equator, the Earth abruptly jumped on its spin axis. A former government geologist, William Hutton, described it like this:

> By examining the motion of the Chandler Wobble for the period between January 1996 and December 2000, it appears that during December 1998 the normal, relatively smooth motion of the wobble path is abruptly deflected; and it takes about 20 months to recover and resume its normal wobble path.

So what apparently happened is that, just as the solar system came into the most perfect alignment with the galactic equator and the center of the galaxy, the supermassive, super spinning and super energetic black hole in Sagittarius A sent out a beam of torsion energy—and the Earth *shuddered*.

Interestingly, as I pointed out, this did not cause any significant seismic events or even minor volcanism. But in the years since, the Chandler wobble has abruptly deflected in December and January just as it did in 1998, and we have had several huge earthquakes during these periods, including the one that caused the Indonesian tsunami in 2004. Now of course, it was reported that *because* of the

Indonesian earthquake, the Earth shifted slightly on its rotational axis.[1] The same thing happened during the Chilean earthquake of 2010 (and probably the similar 2010 quake in Haiti). What I think is that the chicken came before the egg. It was the surge of torsion energy from the galactic center that shook the Earth off its spin axis, and the earthquake was the *result* of that deflection, not the cause.

If that's the case, then we do indeed have reason to be concerned about physical effects from this pulse of energy from the galactic center during the "2012 era" (2010–2023).

But still, though these earthquakes have been devastating to the local inhabitants, the world at large has remained unscathed. Most of the death and destruction have been due to the poor state of preparedness in the countries most affected by the quakes, and to the generally poor quality of building construction in the earthquake zones. Critics will argue that, even if a torsion wave of some kind is really behind these deflections in the Earth's rotational axis, that really doesn't translate to a much broader scale disaster along the lines of surging solar flares or massive, Atlantean scale movements of the Earth's crust. And they'd be right.

Except for one thing: In a hyper-dimensionally connected system, all of the planets have an effect on the physics and their spin energy can be added to the reverberations under certain geometric/astrological conditions. Now, although we are well aware of just how everything lines up in 2012 and beyond, what we don't for certain is if we really have all of the information. What would happen if, as a for instance, we hadn't found all the planets in our solar system yet? What if there were one or two more planets out there that we hadn't accounted for? And where would they have to be to cause a once in 26,000-year-"problem" for us here on Earth?

The answers are in a theory I call the Nemesis Effect.

14
The Nemesis Effect

As we've established, astronomical alignments and eclipses, like the 2012 alignment with the center of the galaxy, have been proven to have an effect on physical instruments and, by implication, our consciousness itself. But the relevance to our present day and the possible transitions we may confront in the next few years may seem tenuous to you at best. Beyond that, the question of whether the measured torsion effects are powerful enough to cause a physical disaster along the lines of films like *2012* and *End of Days* is debatable. Certainly, Hollywood has always made a buck off exploiting the end-of-the-world scenarios, but is there a reason in the physics, the *real* physics we've now established as hyper-dimensional, to be concerned that the doomsayers and the disaster movies may be right? The short answer is yes.

But we can do something about it.

If you've read anything at all about the Mayan calendar or the 2012 end date, you have no doubt discovered that there are as many theories about what will happen as there are sites on the Internet. Everything from magnetic pole shifts to crustal displacements to massive sterilizing solar flares to an alien invasion has been mentioned in connection with the end of the Long Count calendar cycle. And the truth is that in course of the research for *Dark Mission*, Richard Hoagland and I did envision a scenario where some or even all of these events could manifest themselves in this particular time period. And they seemed to center around a pulse of energy coming from the center of our Milky Way galaxy.

Certainly, the Yuga cycles discussed in Chapter 7 seem to have been somehow connected to the solar system's orbit around the galactic nucleus. But what I haven't mentioned so far is that the Mayans also speak of something coming from the center of the galaxy. Calendar researcher John Major Jenkins and symbolist William Henry have both studied the calendar and various related monumental inscriptions. They have both independently come to the conclusion that the *Bolon Yookte* entity (also known as the "nine winds") is actually a reference to the Mayan god Kukulkan, the feathered serpent of the Dresden Codex who unleashes Seven Macaw on the world at the end of the Long Count cycle. They also both cite a passage that claims that Kukulkan will arrive on Earth while riding a "serpent rope" emanating from the center of the galaxy. Both of them have stated that they believe the "serpent rope" reference is related to a theoretical phenomenon called a wormhole, literally a hole in space that connects two distant points through some sort of space-time tunnel, kind of like a cosmic shortcut. This idea is not without merit, but similar to other researchers like author Greg Braden, Jenkins is seeking to interpret the Mayan codices using conventional physics models, and they just don't work. The truth is there is no evidence at all beyond some back-of-the-envelope speculation that wormholes even exist.

But yet we have this consistency. If the Mayans and the Vedas weren't fools (and they demonstrably weren't) then what could be the mechanism of the serpent rope? And what will happen when Kukulkan arrives?

The serpent reference with regard to the center of the galaxy is interesting for a number of reasons. As we learned earlier, most ancient cultures and esoteric societies have a reverence for a symbol called an Ouroboros, a snake in a circular shape consuming its own tail. It is commonly accepted that this image is symbolic of the cyclical nature of time and is inspired by the shape of the Milky Way galaxy itself, which appears as a great circular serpent when viewed from Earth. The Greek philosopher Plato, among others, named this formation the *Suntelia Aion*, the "serpent of light," and predicted that the current age of Man would end catastrophically when some unspecified cycle related to the *Suntelia Aion* concluded.

Image 14.1. Image courtesy of NASA.

So the meaning of the serpent reference, particularly with regard to the center of the galaxy, is a fairly easy one to figure out. Less obvious, unless you've read this book, is what the "rope" part of the "serpent rope" reference might mean. Clearly, the "serpent rope" that Kukulkan will ride is a torsion beam, very similar to the energy that was registered by Hoagland in his Albuquerque experiments in 2006. So, according to the Maya, in 2012 Kukulkan will ride a pulse of torsion energy from the black hole at the center of the galaxy, manifest here in our 3D reality as the *Bolon Yookte* entity, and pass judgment on us all. At that point, I guess depending on what he sees when he gets here, he may or may not resurrect Seven McCaw and call on the great serpent to pour water down from the heavens and wash away the works of Man.

But why, other than the Long Count ending, does this have to be? What is there in the physics that makes 2012 any different than any other year, especially 1998 when the alignments were far closer to perfect than they are now?

The answer is commonly known as *Planet X*.

For more than a century, since Neptune was discovered in 1884, astronomers have been pondering the existence of some kind of planetary body out beyond the orbit of Pluto. Because Neptune and Uranus had some orbital irregularities, it was thought that perhaps there was a large Jupiter-sized planet in the nether regions of the solar system. Recent studies have sought to discount this, and explain the oddities in the outer planets orbits and in the behavior of some comets to natural deviations that do not require an undiscovered planet. In recent years, the term *Planet X* has become associated with everything from the Sun's dead twin to a massive brown dwarf star linked to Zechariah Sitchin's "Nibiru."

Another theory about a far-distant object in the solar system is known as the Nemesis theory. After the publication of the scientific papers we discussed in Chapter 8 that showed a 26-million-year cycle of mass extinctions on our planet, astrophysicists began to consider what could cause such a cycle. They eventually settled on the idea that our solar system is actually a binary (two-star) system, not a single star system. (Gee, isn't that what the Vedas and the precession theorists have been saying all along?) The Sun, they speculated, orbits a dead companion star or a brown dwarf that is way out in the so-called Kuiper belt, a region of space beyond the orbit of Pluto where there is thought to be hundreds of thousands of asteroids and comets and fragments of the stellar nebula from which the Sun condensed. The thinking was that, as this dead star orbited the Sun, it disturbed the objects in a dirty part of the Kuiper belt and sent them on collision courses with the inner solar system (and Earth) on a periodic basis—once every 26 million years. The name they gave this dark star was *Nemesis*.

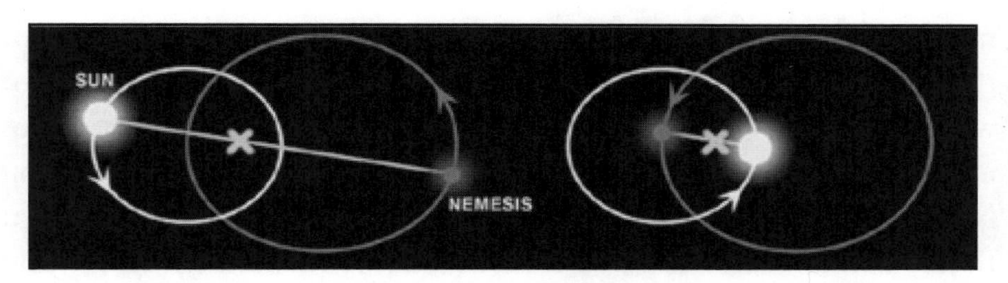

Image 14.2. Theoretical orbital relationship between the Sun and Nemesis. Both stars orbit each other around a common center of gravity. Image courtesy of NASA.

Until the 1980s, neither the Planet X or Nemesis theories had much hope of being tested or confirmed. However, in the early 1970s, two unmanned spacecraft, *Pioneer 10* and *Pioneer 11*, were launched on missions to the outer solar system. Because they made no course corrections of any kind, they could serve as unique test beds to study the effects of gravity at great distances from the Earth. This also made them excellent candidates for data collection in the search for planets beyond the orbit of Pluto.

After more than 40 years, data from both spacecraft show a distinct and unexplainable drag on their acceleration out of the solar system. Various ideas have been put forth to explain away the "Pioneer Effect," including data transmission and collection errors, unaccounted for heat emissions from the probes, or mistakes in the processing of the data. However, after four decades, seven different studies have confirmed that the two *Pioneers* are exiting the solar system much slower than they should be, implying that something is exerting a gravitational drag on them. Faced with a real phenomenon, astrophysicists are confronted with one of two possibilities: Either the laws of physics are different outside the solar system (a distinct possibility), or there is something unknown out there beyond the depths of space that is tugging on them with its gravity and holding them back. Because the first idea, that maybe the speed of light is different outside the orbit of Pluto, is deeply troubling to scientists, most of them argue either that the observations are simply wrong or that there is something big out there pulling on the two *Pioneers*. For want of a better name, they usually call it Planet X. But if there was a Planet X out there, it would be also measurably affecting the orbit Neptune and Uranus, and it demonstrably isn't. So the best current theory is that some kind of interaction with "vacuum energy"—that is, zero-point energy—is slowing the spacecraft down. What they haven't considered of course, because their paradigm does not allow it, is that a variation of both ideas is responsible for the effect.

What Richard C. Hoagland and I proposed in Chapter 2 of *Dark Mission* was that there is at least one or two undiscovered planets way out beyond anything that could be considered part of the solar system. And these planets are generating torsion waves that are impeding the escape of the two spacecraft from the solar system. It is also possible that the laws of physics might actually be different beyond the orbit of Pluto. After all, as we've established in earlier chapters, if the geometric configurations of the known planets in their orbits can affect the laws of physics inside the solar system, then it stands to reason that the laws of physics might be considerably different outside their direct influence. That might affect known "constants" of physics, such as the speed of light. If in fact the speed of light is slightly faster beyond the orbit of Pluto, then the radio signals from the two *Pioneer* spacecraft would arrive back on Earth faster than expected. This would create the illusion that they are closer to Earth than they actually are, because scientists use the timing of the signal transmissions (which are assumed to travel at the speed of light) to determine the distance of the spacecraft from Earth.

Whatever the case, we stated unequivocally that in our model, there has to be at least one massive Planet X out there somewhere, or two smaller bodies in retrograde (backward) orbits. Our theory is based on van Flandern's projections from the stellar fission theory, and once again is linked not to gravity or magnetism, like conventional physics, but to spin energy. Richard was the first to recognize that a system's total spin energy and its total luminosity (brightness in all ranges of the electromagnetic spectrum) were connected. If you look at Jupiter and its nine moons as a system, and you add up all of the spin energy of all of them together, you get a set value for the entire Jovian system. What he found was that the more total system spin energy a planet and its moons had, the brighter it got. This is completely in line with the hyper-dimensional argument that energy is actually generated by spin.

When he plotted all the planets and all the moons on a graph, it followed a curve that was predictable and well within the expectations for his idea. But there was one notable exception: the Sun.

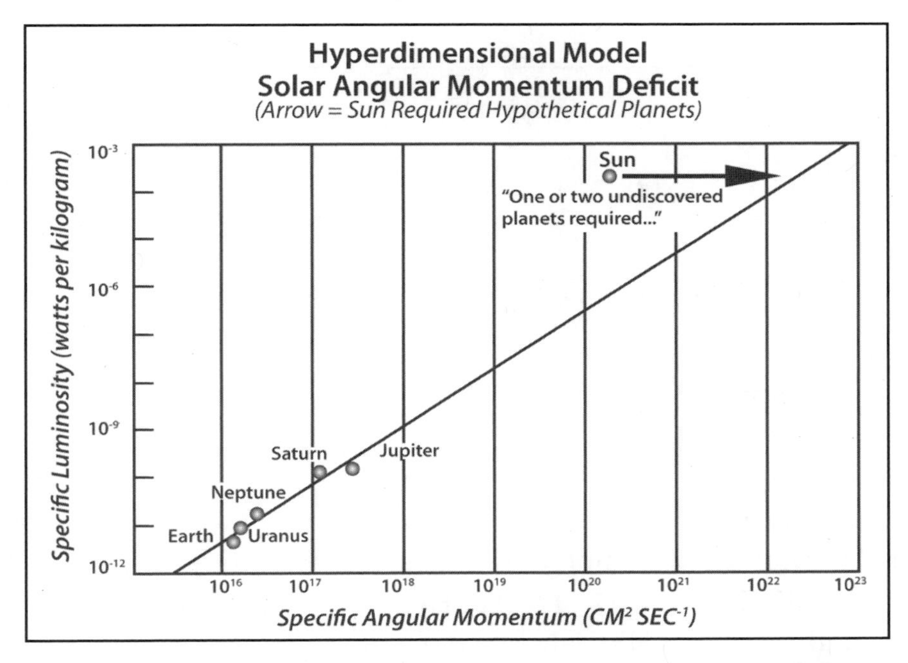

Image 14.3. Luminosity vs. spin energy. One or two undiscovered planets beyond the orbit of Neptune would move the Sun to the right and place it on the predicted curve along with all the major planets. Image by Krys Lilly.

As far as the planets go, the graph is eight for eight. Yet when we add up the spin energy from all the planets and all their moons and put them all on the graph, the Sun should be right where the rest of planets are, but it isn't. For some reason, the Sun was inexplicably off the curve. It was in fact far brighter (more energetic) than it should be if the theory was correct, given its spin energy. So either we're wrong, or...There are some planets we haven't discovered yet.

Now, you can certainly chide me for doing what I've complained about scientists in other disciplines doing earlier in the book (namely moving the goal posts to make my theory work). But don't forget: The difference here is that we are already eight for eight with the other planets, and we are making a very specific proposal that can prove us right or wrong very easily. And there's more.

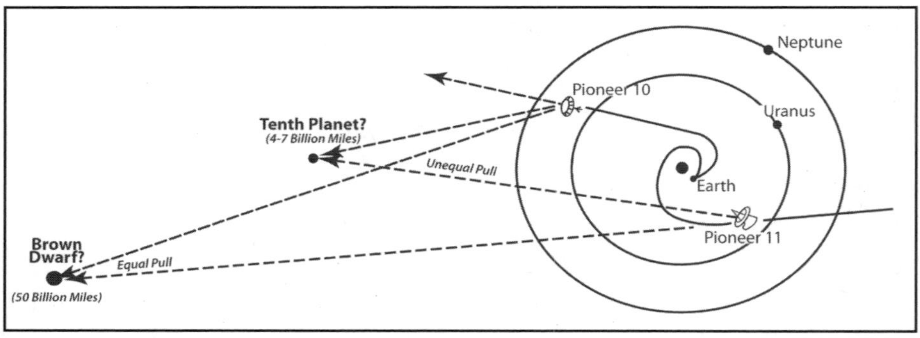

Image 14.4. Image courtesy of NASA.

There is some evidence that NASA was already pretty certain there were a couple of undiscovered planets out in the nether regions of the solar system when they launched the *Pioneers*. In a 1982 *Science Digest* article, Dr. J. Allen Hynek, a well-respected astronomer and sometimes UFO investigator, wrote about the search for the Planet X's and how the *Pioneers* could be used to search for them. In the article, he included a diagram that showed the two objects relative to the *Pioneers*' paths and the known solar system. The first body was shown to be 4 billion to 7 billion miles out, and was labeled the "Tenth Planet." The second object was labeled "brown dwarf?" The brown dwarf was listed as being 50 billion miles out.

This is a mind boggling distance. Neptune, the farthest planet in the known solar system, is about 2.7 billion miles from the Sun, or about 30.1 Astronomical Units (93 million miles, the distance from the Earth to the Sun). At 50 billion miles out (550 A.U.s), this hypothetical brown dwarf or dead star would be so far away it would be really stretching it to regard it as part of our solar system. In fact, one of the major arguments against Sitchin's Nibiru theory is that its assumed 3,600-year elliptical orbit would be unstable. But a 50-billion-mile orbit (again, 550 A.U.s!) works out to something like a 13,000-year orbital period, if the orbit is roughly circular. That once again is considered to be just too far away to be gravitationally tethered to the Sun and to be able to sustain a stable orbit.

However, as I have said repeatedly in this book, gravity is only a weak force in a much bigger, hyper-dimensional universe, and there is no reason to think that the same unseen force (torsion) that holds everything together isn't also keeping this object linked to our Sun and the solar system. Besides, it's now established that such an orbit is theoretically possible. In 2009, astronomers at Toronto University took an infrared (heat energy) image of a very young (5 million years old) star. They also spotted a very hot, very dense object orbiting the star at a distance of 330 A.U.s from its parent. Obviously, they had inadvertently taken the first photograph of a "Hot Jupiter" that had been ejected from its young parent star fairly recently and was probably still spiraling away from it. But given that the distance is some 60 percent of the projected 50-billion-mile orbit of the "brown dwarf," there is no reason to think that such an object could not be that far out, orbiting our Sun, and still be part of our solar system.

What we don't know—what we cannot know unless they decide to tell us—is whether NASA or any other space agency has actually found either Planet X or the brown dwarf companion to our Sun. But I can tell you this: They are looking for it.

In Dr. Hynek's 1982 *Science Digest* article, he mentioned that if either object were out there, the (then) upcoming Infrared Astronomy Satellite (IRAS) could be used to detect them. In fact, he made it clear that they intended to do so. A few months after IRAS was launched, stories began to appear in the press regarding an unusual object that was "as big as Jupiter" but so cold it didn't shine at all in the visible light spectrum. It was spotted, according to Dr. Gerry Neugebauer of NASA's Jet Propulsion Laboratory, in the constellation of Orion and it was at a distance of "50 billion miles." When I wrote to Dr. Neugebauer asking him if there had been any follow-up observations after the big press splash, he denied knowing anything about any follow-ups or even remembering the object in question. Now personally, if I was quoted on the front page of the *Washington Post*, I think I'd remember that one for a little while, but hey...

Today, Wikipedia claims that this object was later found to be merely a distant cold galaxy that was mistaken for local object, but I have looked at the papers in question and have yet to be convinced that this is truly the case. And Dr. Neugebauer's implausible denial makes me think that it was certainly not just some distant galaxy that he and others mistook for a brown dwarf star.

The point is this object, whatever it was, fits the textbook description of a brown dwarf, not to mention that it certainly fits the Vedic descriptions of the dark star around which the Sun revolves. And it also fits the Nemesis theory quite well, for that matter. So if there was some kind of NASA cover-up on this, after initially making an announcement about it, why would they do that? Beyond confirming our own theory as to the link between spin energy and brightness, what would be the point in keeping such a monumental discovery quiet?

The truth is, there wouldn't be one—unless the hyper-dimensional model of the Universe is correct.

If Nemesis and/or Planet X truly exist, and I think it's a reasonable bet they do, then they are well and truly part of our solar system. This must mean that they have spin energy like all the other planets, and if they are enormous gas giants like Jupiter, Saturn, Neptune, and Uranus, they are likely, because they were ejected from the Sun earlier in the Sun's life cycle, to have a great deal more spin energy than even Jupiter does. So if the hyper-dimensional model is correct, then that means that these two objects, let's call them Nemesis and Nibiru just for fun, stand to have a great deal of influence on the physics of our planet. Maybe even more than the other more familiar planets do.

If Nemesis is out there, some 50 billion miles away in its 13,000-year orbit, then that means that we must include it in our assessments of the 2012 to 2023 end-time scenarios. As we learned in Chapter 8, when the planets are in certain geometric alignments along their orbits relative to the Sun or Earth, it has an effect on the physics here. Those alignments corresponded to the same configurations in astrology, the squares, trines, sextiles, and oppositions. As we look at the 2010–2016 period coming up, we have the alignment of the Earth, Sun, and the center of the galaxy. So depending on where Nemesis and/or Nibiru are, they must also be taken into account when looking at the possible physical effects of this alignment. I find it interesting that 50 billion miles works out to a roughly 13,000-year orbital period for Nemesis, because that is just about half a precessional cycle of (roughly) 26,000 years. Many scholars and researchers, like author Graham Hancock, argue that the Great Flood of the Bible took place 12,500 years ago, or just about half of a 26,000-year precessional cycle. They further argue that the Egyptians marked this time with the building of Pyramids and Sphinx, and that they were meant to mark that period as the time of the

Great Flood, the time that wiped out the advanced civilization the Egyptians call the *Zep Tepi*, "The First Time." Hancock points out that this is the reason the Egyptians keep track of the precession of the equinoxes through the Great Pyramid of Giza, so that they will know when we have come one half precessional cycle around from that last devastating flood.

The question is: why? If the Flood was just a random event, caused by a comet impact or some other natural calamity, there would be no point in tracking the movements of the heavens. But if they understood that these catastrophes were cyclical, as they understood time to be, then it would make all kinds of sense for them to know when the next danger point would be so they could be prepared for it as best they could.

I am making a further assumption here. I am going to assert that the Egyptians, along with the Vedas and Hopi and all the other ancient civilizations that seemed to obsessively track precession, understood that these catastrophes were cyclical, and that they were somehow driven by the alignment of the Dark Star Nemesis with the center of the galaxy. The mechanism for this fits very well with all that we know about torsion and hyper-dimensional physics.

If Nemesis is really out there, and if it is on a 13,000-year orbit, and if it is coming into alignment along with the Earth, the Sun, and the center of our galaxy during this 2010–2023 period, then it would have a substantially amplifying effect on all of the torsion energy, the serpent rope, that is generated from the black hole at the heart of our galaxy. If the last such alignment was in 10,500 BC, 12,500 years ago, then we are certainly in the ballpark for another period of great change and upheaval. Not only would adding the torsion energy to the 2012 equation have a potentially strong effect on our planet, it could and should have an equally powerful effect on our consciousness.

So this may be why this time period—the post-2000-era—was so important to the Vedas and the Mayans and Egyptians and the Hopi. They were trying to warn us to get ready. The question is: get ready for what?

Well, that very much depends on where Nemesis is exactly in its orbital path. Remember that RCA's John Nelson found a correlation between the traditional astrological aspects and the Sun's magnetic output. If Nemesis is between the Sun and the galactic center, then this will be essentially a conjunction. In astrology, a conjunction is an aspect that amplifies other aspects (planetary alignments) in the astrological chart. That means that depending on the positions of the other planets, the effect, good or bad, will only be *amplified* by the conjunction of Nemesis with the galactic center. However, if Nemesis was at the other end of the celestial sphere, say in the constellation of Orion, which is 180 degrees from Sagittarius, then this would be an astrological *opposition*, a very stressful but not necessarily negative aspect. An opposition will tend to exaggerate the differences between the other planetary alignments, whereas the conjunction tends to unify them. Now, as we just learned, the most likely location for Nemesis is based on the 1982 IRAS discovery, which found it to be in Orion, in effect creating an opposition. If that is the case, then what is going to happen during the 2012 era will depend in large part on where the other planets are in relation to the Sun and the Earth.

The 2012 Astrological Chart of the Earth

Obviously, plenty of astrologers have looked at the Earth's astrological chart for December 21, 2012. Quite frankly, from what I've read in my research, there is no consensus as to the meaning of it, and none of the charts take into account the possibility of a Nemesis star or Planet X, nor do they understand the underlying physics involved as we've learned in this book. But there do appear to be two astrological configurations that argue that December 21, 2012, will be a nexus point or at least the beginning of a significant transition for humanity. The first significant aspect is called a Yod, also known among astrologers as "the finger of God."

A Yod is an arrowhead-like alignment in an astrological chart that signifies two planets in opposition to a third. The three planets involved in the December 21, 2012 Yod are Pluto, Saturn, and Jupiter. Pluto represents radical transformation, death and rebirth in astrology, and Saturn represents the Earthly realm and all matters practical and adherence to social structures. In the 2012 Yod, they are in opposition to Jupiter, which rules over the principals of growth, expansion of thought, and life.

The other interesting configuration is called a T-Square, and involves Venus, Jupiter, and Neptune. Jupiter and Neptune are at 90 degrees to each other, as are Neptune and Venus. This makes Neptune, the planet of spirituality, mysticism, psychic phenomena, ascension, and floods the focal point of the alignment. Venus, the planet of sex, romance, sensuality, comfort, ease, and the material, is in direct opposition to Jupiter.

It isn't too difficult to interpret what this all means. The chart indicates that on December 21, 2012, our consciousness will be in a state of conflict. We will face many choices about our philosophical and religious beliefs, our social systems, and our value structures. Most likely, our monetary systems and our forms of government will be under scrutiny. There will be a growing disconnect and a clash between the desires of the flesh and of the spirit.

We will be faced with a Choice about our future. One path will lead toward Pluto, meaning death and rebirth. One path will lead toward the influence of Jupiter, which means we will be able to expand our lives and our thinking to include new ideas and perhaps entirely new ways of living with each other based on moral and spiritual values instead of greed and material desire. But each of us, and our planet itself, will be confronted by this tension, this pressure of the oppositions. If Nemesis is in fact out there, whether it is in Orion or Sagittarius, it will amplify the tension of this chart, and force us to make a Choice.

I don't want to alarm you too much about the physical nature of this transformation, but we must consider it. If the Great Flood of the Bible is indeed linked to the last alignment of Nemesis with the galactic core, then that certainly could have been a contributing physical factor in a crustal displacement or magnetic pole shift. It's also important to understand that in the physics, it doesn't matter that the Sun was not conjunct with the center of the galaxy as it will be in 2012. If we are right about Nemesis's orbital parameters, then in 10,500 BC Nemesis and the Sun were conjunct somewhere in Orion, and the galactic center would have been 180 degrees away, in opposition to that unifying force. So if the Earth did experience a pole shift of some kind, astrologically and hyper-dimensionally, 2012 is a very different scenario. This time, we appear to have a Choice.

Remember that the Hopi told us that in the previous world, the very technologically advanced third world of Man, the people had completely lost sight of spirit and were consumed by desires of the flesh and material. This world is far different, with many the world over still holding their hands out to God and seeking something higher. It also should be noted that Neptune, which plays a major role in the 2012 Earth chart, rules over floods. But remember also that water is frequently associated with knowledge being poured into our consciousness. So the flood of Neptune, like the serpent pouring water down upon the Earth from the last frame of the Dresden Codex, may be referring to a symbolic flood of knowledge and enlightenment, rather than a literal flood caused by a physical pole shift.

What I think this all adds up to, and there are many that agree with me, is that in these next few years we will be forced by circumstances to make a Choice: to either stay in the lower realms of earthly desire or to seek out a new way, a higher path of service to others and spiritual enlightenment. If the majority of us decide to stay in the intellectual/material world that we have created, then all of the prophecies and all of the alignments say we are in for a stressful, difficult, and possibly cataclysmic time of it. But if we instead Choose the way of the light, then we can have an entirely new world—one where reality itself may be altered for the better. But as I will show you next, it is all up to us. We are not helpless victims of this process. We are active participants in it. We have all come here at this time to decide what we want. Not for ourselves, as The Secret tempts us to do, but for our world.

The question is: Which path will you choose?

15

The Doomsday Scenarios

No discussion of 2012, the Mayan calendar, or even the post-millennial period in America would be complete without some recognition of the various doomsday scenarios that have been bandied about the internet and ruthlessly exploited in various 2012-themed television specials. I was loath to go here in this book, primarily because I simply don't believe in any of these end-times scenarios, for reasons I hope to prove to you. Certainly, there is evidence that all of these events have happened at one time or another in the history of our planet, and there is some reason to think these types of events are cyclical. But that said, what I know, and what I hope to show you, is that, though all of these traumatic Earth changes are possible, they are not inevitable, or even likely. What happens is up to us.

Transformation and Revelation

The word *apocalypse* comes from the Greek *Apokálypsis,* meaning "revelation" or "lifting of the veil." It can also be seen as a new beginning rather than an end of something else. The end of the world as we know it may not necessarily correlate to the end of the world period. Apocalypse, however, is in many ways the perfect theme for our time. As we move from the occult (hidden) Age of Pisces to the enlightened Age of Aquarius, we are supposedly going to come to know all things that have been hidden from us by our institutions and religions. Obviously, if such revelations are dictated by the physics of consciousness, then the anger and backlash of all the things we have been lied to about could certainly topple long established institutions and even our form of government.

That said, there are many, some of whom I respect, who argue that the apocalypse will be of a physical nature—that there will be a great culling of the human race and judgment in the end of times. The Hopi speak in their prophecies of the Great Cleansing of the Earth, as was done in the previous three Worlds of Man. The Vedas speak of the end of the Kali Yuga cycle just before the dawn of the new Golden Age, and how Kali will come to Earth and destroy all of the wicked and unfaithful.

Revelation 20:12 in the Christian Bible says:

> "And I saw the dead, the great and the small, standing before the throne; and books were opened: and another book was opened, which is the book of life: and the dead were judged out of the things which were written in the books, according to their works."

Even the Mayan Prophecies related in Chapter 9 tell of each of us facing a Hall of Mirrors in "the time without time," when we must confront our own judgments and behaviors.

All these sources are really remarkably consistent. They each tell of a revelation, or apocalypse of some kind, after which we all must apparently take sides and then something really bad is supposed to happen that will separate the good from the wicked. For centuries, our cultures and religions have been fascinated by what that "something bad" is going to be. The possibilities generally fall into a few main categories: something called a "Z-Pinch" solar flare, a magnetic pole shift, a full on crustal pole shift, or possibly a meteor or comet impact. Let's take a look at each of these in turn.

The "Z-Pinch" Solar Flare

This idea comes from the notion that the Sun will somehow get excited to the point that it unexpectedly flares up and sends out an enormous coronal mass ejection that basically fries the planet. Because of oddities in the Earth's magnetic field, this flare will be drawn down through the poles, the weakest part of the field, which normally protects us from such stellar eruptions. What makes the Z-Pinch scenario unique is that under certain exotic conditions, the particles in the magnetic field of the Earth can become ionized and exited into a state called a plasma, and the superheated gas would be captured by the magnetic field and could scorch the Earth for days or even months at a time.

The idea was primarily put forth by physicist Dr. Paul LaViolette in the 1980s. He believed that the center of the galaxy periodically "exploded," sending out bursts of gamma ray energy he called the "Galactic superwave." Although Earth's magnetic field would normally protect us from such a distant burst of radiation, LaViolette also speculated that the local area outside our solar system was filled with ice and dust, and that these bursts of energy pushed large quantities of interstellar dust into the interior of the solar system, causing the sun to "choke" on it and begin to flare up. He later argued that such an occurrence 12–13,000 years ago was the reason that last ice age ended, and that certain Mayan prophecies concluded that the cycle would repeat itself in the 2012 era.

There has been some support for his ideas. On the *Apollo 11* mission, the first to land humans on the surface of the Moon, Neil Armstrong and Buzz Aldrin made observations of what they called "glazed donuts" on the Moon. A 1969 scientific paper by the brilliant Dr. Thomas Gold concluded that the glazed donuts were clumps of moon dust that had been subjected to rapid and high intensity heating, resulting in a glass layer covering the clumps of Moon dirt.[1] Gold placed the dating as sometime in the previous 30,000 years, very recently in astrophysical terms. In order to create the glazed donuts on the surface of the Moon, the Sun would have had to emit on the order of 100 times its normal heat-energy output for a period of between 10 and 100 seconds. Such an event, hitting the unprotected Earth, would have scorched the planet, possibly boiled many of the oceans, and caused extinctions on a mass scale. Whether or not the Z-Pinch effect would have grabbed the solar flare and held it over our heads for months at a time is debatable. Gold did conclude, however, that the Sun periodically flared up in this manner, perhaps every 10–15,000 years or so. Gold also speculated that the flare could have been caused by a large comet or asteroid randomly hitting the Sun, which caused it to spew out a major flare. Either way, the results aren't likely to be pleasant if it happens.

Supporters of the Z-Pinch solar flare scenario point to various monumental inscriptions, like the Great Cross of Hendaye, in Hendaye, France, as a sign that this scenario is somehow tied to the 2012 period. They also are fond of pointing out numerous ancient cave paintings, sometimes dated to around 10,500 BC, which depict people, plants, animals, and the skies seemingly on fire. Jay Weidner, an author and Cross researcher, believes that the Cross was built by a mysterious French alchemist named Fulcanelli (likely a pseudonym) and that it contains an alchemical code that argues that the Earth and the wicked will be extinguished by fire from the Sun in 2012.

Most of the advocates of this scenario have pointed to the absence of solar activity and sunspots during the current solar minimum as a sign that the Sun is getting ready for something big. The Sun's activity waxes and wanes in cycles that take about 11 years to complete. Periods of low activity are characterized by a lack of sunspots and generally lower temperatures here on Earth. Periods of higher activity mean there are lots of sunspots, magnetic storms, solar flares, and higher temperatures on Earth.[2] The recent solar minimum was one of the calmest on record, leading to widespread speculation that the 2012 solar maximum would be much worse than normal. However, we are now well into the ramping-up phase of the current solar cycle, and most solar physicists now say this will be a very mild run-up to the next solar maximum.

The other flaw in all this is that our magnetic field does indeed protect us from such solar outbursts, and though there may have been a time in the past in which something like the Z-Pinch occurred, that doesn't mean it will happen anytime soon, much less in the 2012 transitional period. And even if the last such flare was timed to the estimated time of the Great Flood (10,500 BC), that was a flood, not a flare. Now, it can be argued that *something* had to end the last ice age around 13,000 years ago, and a flare that was captured in a Z-Pinch type of scenario could certainly have raised global temperatures and melted the ice. This also could have conceivably led to catastrophic flooding around the world, particularly in coastal zones. But if that's the case, why didn't the ice return when the Z-Pinch flare subsided? Unless the Earth was also knocked off its spin axis or the Sun was at an all-time high in terms of its energy output, then presumably the northern hemisphere should have frozen up again and the glaciers should have returned. Instead, the polar ice caps continue to melt. So I think we can pretty much discount the massive solar flare model for the moment.

Also, as we'll learn later, there are some steps that can be taken to guard against this specific scenario using currently available technology. Personally, for those reasons, I put the solar flare Z-Pinch scenario near the bottom of my end of days list.

Magnetic Pole Shift/Crustal Displacement

This one is a favorite of most of the 2012 end time advocates, primarily because there is some evidence it has happened in the past. The concept of geomagnetic reversal, or a magnetic pole shift where the Earth's magnetic field jumps from a northern polar orientation to a southern one or vice versa, was first proposed in 1920s Japanese geophysicist Motonori Matuyama when he found undersea volcanic rocks that showed evidence of different magnetic pole orientations. When molten rocks cool, the metals contained in them are magnetized, and these magnetic properties are "watermarked" with the magnetic field orientation of the Earth at that moment. As such, they represent a snapshot in time, at least as far as the orientation of the Earth's magnetic field in concerned. At first, the notion that the Earth's magnetic field could be reversed at all was ridiculed in science, but as more and more magnetized rocks were discovered and studied it became obvious that the Earth's magnetic poles had indeed shifted back and forth repeatedly throughout its history. Most of these changes had been in the far distant past, and many such shifts took place in clusters. (About 15 million years ago, the field reversed itself 51 times in a 12 million year period, about once every 235,000 years.) These clusters were counterbalanced by "super-chrons," long periods where few if any re-orientations occurred. The causes of the magnetic pole reversals are not understood at all. Some argue that they are due to internal forces related to the Earth's magnetic "dynamo," its spinning, magnetic molten core. Others, like the previously mentioned Dr. Paul LaViolette, argue that they are due to external influences, like his Galactic superwave.

The reversals themselves had been thought to be random, but research by a team of geophysicists in 2006 found that they did indeed conform to a mathematical pattern, which implied the reversals are linked rather than independent of each other.[3] The data also indicate that they happen very slowly, over periods of hundreds of thousands of years at least, and the last one was about 800,000 years ago. This means that we could be due for one soon. Or, we could already be in the process of it.

As far as the effects of a geomagnetic pole reversal on human beings, unless they were linked to a Z-Pinch event, or a comet or meteor impact, they would appear to be minimal. We humans have survived at least one in our 2-million-year history, and the only likely effect might be on the weather. Storms and winds could increase substantially, but probably not enough to cause much damage beyond the coastal areas. Electronic instruments and navigation systems could be in for a much worse time of it though, and we are becoming increasingly dependent on both for our everyday survival, right up to the delivery of goods to markets. A sudden pole reversal might render most computerized systems impotent, and that could lead to a breakdown in supply and distribution, resulting in widespread famine and unrest because of it. Beyond that, a geomagnetic pole reversal wouldn't be that much of a threat to the entire human race, especially if you consider that no one really knows if computer systems would be adversely effected or not.

The Crustal Displacement Theory

A related concern, crustal displacement, or a sudden geographic spin axis shift, would be a whole different story, however. The concept is that, under certain kinds of geologic stresses, the upper crust of the Earth can break away and slide over the inner mantle like the loose skin of an orange. It is this idea that is at the core of the plot of the 2009 movie *2012*. As a result of such a slippage, continents would rise and fall along the lines of the Atlantis legends, the oceans would slosh over the coastlines and wash away the works of Man, and volcanoes and earthquakes would devastate the planet. After such a quick shift, climates would change dramatically as a new spin axis was established, and the few survivors of such a cataclysm would have a pretty rough time of it getting civilization started again.

The idea of a pole shift related to crustal displacement was first proposed in the early 20th century by Charles H. Hapgood (1904–1982), a Harvard-educated historian with a PhD who dabbled in various scientific disciplines. During World War II, Hapgood worked for the OSS (Office of Special Services), the precursor agency to the CIA. After returning to academic life after the war, he became interested in the myths and legends of the ancient world and especially that of the lost continent of Atlantis. Searching for a mechanism that could explain the collapse of an entire continent into the sea, he settled on the crustal displacement theory as the most likely idea and published it in a 1958 book entitled *The Earth's Shifting Crust*. The idea came at time shortly after the publication of Emmanuel Velikovsky's catastrophist epic, *World's in Collision*, and it also caught the attention of a public hungry for more fresh ideas on the nature and history of the Earth. The ideas that Hapgood put forth in *The Earth's Shifting Crust* were so engaging that Albert Einstein himself wrote the foreword to the book shortly before his death:

> I frequently receive communications from people who wish to consult me concerning their unpublished ideas. It goes without saying that these ideas are very seldom possessed of scientific validity. The very first communication, however, that I received from Mr. Hapgood electrified me. His idea is original, of great simplicity, and if it continues to prove itself of great importance to everything that is related to the history of the earth's surface. I think that this rather astonishing, even fascinating, idea deserves the serious attention of anyone who concerns himself with the theory of the Earth's development.

The "electrifying" idea that Hapgood asserted in *The Earth's Shifting Crust* was that the physical poles, the alignment of the Earth's spin axis, had radically and swiftly shifted at least three times in the last 100,000 years, the most recent being completed about 10,500 BC (yes, right in the same time period that Hancock and others place the era of the Great Flood of the Bible). He cited a number of factors in his theory, namely that the crust can slide on a somewhat liquid-like layer between it and the hard inner mantle, and evidence that the magnetic poles had shifted frequently as well.

The mechanism cited by Hapgood was the accumulation of ice at one or both of the poles. Eventually, after thousands of years, too much ice at one location would cause an imbalance in the "moment of inertia tensor." Basically, this is the center of gravity of the Earth's spin axis. Hapgood visualized that the crust would then slip over the outer mantle, while the inner core would retain its spin axis orientation. In other words, the Earth wouldn't "flip" its spin axis, as in the Indonesian tsunami of 2004. Rather, the crust would just slide loosely over the inner core. Einstein commented on this as well:

> In a polar region there is a continual deposition of ice, which is not symmetrically distributed about the pole. The Earth's rotation acts on these unsymmetrically deposited masses [of ice], and produces centrifugal momentum that is transmitted to the rigid crust of the Earth. The constantly increasing centrifugal momentum produced in this way will, when it has reached a certain point, produce a movement of the Earth's crust over the rest of the Earth's body, and this will displace the Polar Regions toward the equator.

However, as the years went by, the evidence began to accumulate against Hapgood's idea. Einstein himself did some calculations and pointed out that the polar ice wouldn't weigh enough to cause such a catastrophic slippage, and Hapgood eventually agreed. Also at the time Hapgood published *The Earth's Shifting Crust*, the idea of plate tectonics hadn't taken hold yet, and the Earth's crust was thought to one single piece. This would make the magnetic rock data more questionable, because the magnetic field polarity fluctuations were likely to be affected by local conditions. By the time he wrote another book on the subject, *The Path of the Pole,* in 1970, Hapgood had backed off his sudden\cataclysmic model to a more gradual one, saying that he believed the physical poles shifted over roughly 5,000-year periods followed by 20,000- to 30,000-year periods of stability.

But without ice as the primary causal mechanism, his idea began to lose momentum because another acceptable mechanism failed to materialize.

Hapgood then wrote that asteroid impacts or some undefined "underground" process must be the cause, but was never really able to assign a plausible cause to his model before his death in 1982. However, in the mid-1970s, his crustal displacement theory got a significant boost from another highly controversial idea: Zechariah Sitchin's "Nibiru."

The 12th Planet

Also known as the "12th planet," Nibiru is at the center of one of the most frightening end-time scenarios. Supposedly on a long elliptical orbit that brings it through the inner solar system every 3,600 years, Sitichin's theory is that when Nibiru passes close to the Earth, the two planets' magnetic cores try to align like two magnets and the Earth's outer crust is torn loose from its inner molten core, causing something of a really bad hair day for everybody.

Sitchin was drawn to the idea of a 10th planet in our solar system by his study of the genesis myths of ancient Sumer, the oldest known human civilization, dating back to 5500 BC. Having taught himself to read cuneiform, their ancient symbolic written language, Sitchin interpreted the texts as telling the tale of a mysterious "12th planet" (they counted the Moon and Pluto as planets) that passes through the inner solar system once every 3,600 years. He claimed that the planet, which he says the Sumerians called Nibiru (after Neb-Heru, "the place [of the] crossing"), is on an elongated orbit that takes 3,600 years to complete. In his version of Sumerian cosmology, Nibiru was inhabited by a race of beings called the Anunaki, who came to Earth more than 400,000 years ago. Landing in what is now modern-day Iraq, they stayed for several hundred thousand years, during which time they genetically engineered the human race by splicing their own DNA with that of the then indigenous Neanderthal humans they found on Earth ("We shall make Man in our own image"). The result was Cro-Magnon (modern) man.

According to a further reading of even more ancient scrolls, Sitchin concluded that the Great Flood of the Bible was caused by an extremely close passage of Nibiru to the Earth about 12,500 years ago. When it got too close to Earth, the magnetic core of the much more massive Nibiru created a tug on the nickel-iron magnetic core of the Earth, and like two magnets of opposite charge the Earth's core swapped to align with that of Nibiru and the entire planet toppled over on its axis. The results were pretty much as I've described previously, and supposedly most of humanity was wiped out in the process, with only the Sumerian version of Noah and a few humans surviving.

Again, Sitchin's account is entertaining at the least, but there is little to back it up. Cylinder seals that he claims show that the ancient Sumerian's knew about the existence of Neptune and Pluto (which could not be found without a telescope) in addition to Nibiru are controversial at the least, and there doesn't seem to be any correlative evidence in the geologic record of a geographic pole shift. According to most geologists today, the phenomenon of true polar wander is a reality, but strata and ice core records indicate that although the spin axis poles of the Earth do shift, they do so very slowly. A 2001 paper confirmed that the axis of the Earth has not shifted by more than 5 degrees over the last 130 million years, and even then it does so at the nearly imperceptible rate of 1 degree every 1 million years.

So much for Nibiru.

So if the geologists are right, the Earth isn't going to tip over on its axis anytime soon, nor are we going to be engulfed in some massive solar flare that will consume us all in some hellish firestorm. But I can tell you this: There do appear to be some people in governments all over the world who are worried about pretty much everything to do with 2012.

The World in Peril

In my former life as a born-again conspiracy theorist, I got pretty good at reading between the lines and understanding the symbolic language of governments and esoteric societies, like the Freemasons. In looking back at the history of catastrophe scenarios like Velikovsky's and Hapgood's, I can't help but notice that both went through Albert Einstein in a sense in that he was connected to both men, if only by professional association. So as I began to look at the various 2012 end-time scenarios that were in the popular culture, I wondered if perhaps there was some other connection, something that would tie it all together for me and give me an indication which way the various governments of the world were thinking, and what, if anything, they were doing about it.

It didn't take long to find it, courtesy of my old friend and mentor, Richard C. Hoagland.

It turns out that Richard had been talking about a very interesting book for more than a decade. It had come to him apparently through his girlfriend, Dr. Robin Falkov, and our mutual friend Sean David Morton. The title of the book seemed as ominous as it was preordained: *World in Peril*.

At first glance, that seemed an inappropriate title. The subtitle was oddly mundane by comparison: "*The Origin, Mission & Scientific Findings of the 46th/72nd Reconnaissance Squadron*." Hmm. The scientific findings of a *recon* squadron? Not exactly the stuff of eschatological legend. It turned out the book was something of a memoir written by Ken White about his father's reminiscences of his Army days with the recon squadron right after World War II. The 46th (later the 72nd) recon squadron was the first operational unit under the Strategic Air Command (SAC), a division of the United States Army Air Force. Just after the war, White's father, Major Maynard E. White, was given command of the squadron. The squadron was assigned to Project Nanook, a scientific research project that (in part) was ordered to map the actual locations of both the physical North Pole and the magnetic north pole. Because SAC was created to form a strategic nuclear bombing force for the United States, it was part of their mandate to figure out how to guide aircraft over the North Pole. With Soviet Russia as the likely enemy in the rapidly heating up Cold War, navigating over the pole to strike at Russia was an imperative. Using a variety of sensitive instruments, the squadron flew numerous missions over the poles until enough data had been gathered, at which time it was delivered by Major White to his superiors and a team of scientists working with the Pentagon. The whole book is fairly innocuous until you reach the chapter entitled "The Flip of the Earth."

What the squadron apparently found is that the both of the Earth's poles—the physical and magnetic—are indeed slipping, and at an alarming rate. Major White told his son of meetings he attended at the Pentagon concerning this specific subject:

> The present interglacial age—the Holocene Epoch—began about 10,000 years ago… and can be expected to end within the next 2000 years. This estimate correlates very closely with that of the government scientists, who predicted the next 'flip' of the earth could occur any time between 17 and 1000 years of the date of their study, which was conducted in 1947.

The "interglacial age" is the age we now live in, the time since the last great ice age ended about 12,000 years ago. As the ice caps and glaciers receded, more land became habitable and accessible to humans for farming and development. No one knows exactly why the previous epoch, the Pleistocene epoch, which lasted some 2.5 million years, ended. However, given that ice sheets once extended as far south as Washington state in the United States, a movement of the spin axis pole is a distinct possibility.

The book doesn't make it clear exactly what information the scientists had to go on that led them to conclude that the poles would "flip" radically (as opposed to a slow gentle drift) in the next 17 to 1,000 years. But it does say that they argued about how to best inform the public of what they had discovered.

> At one of the scientific meetings that Major White attended in the Pentagon in early 1948, the scientists discussed the advisability of alerting the public to the pending polar-flip phenomenon. None of the scientists would agree to withhold the information from the public; but, on the other hand, neither could they agree on how to release it. The knowledge of this phenomenon, some felt, could in itself destroy the moral fiber of society....[4]

If you've read *Dark Mission*, you'll note that these fears are very similar to those expressed by the authors of the Brookings Report, which detailed how best to inform the public in the event that NASA discovered extraterrestrial artifacts on the Moon or Mars. But it also says that most of those fears were allayed when stories about the pending pole shift were circulated in newspapers and magazines, and the public had no great reaction to it.

Now, call me suspicious, but it seems to me that someone like Albert Einstein, the leading scientific genius of his day, would have most likely been called in on this as a consultant. After all, he had numerous security clearances, and it seems almost unthinkable that he wouldn't have known *something* about these studies, especially if they were perceived as having a great deal to do with the possible future survival of humanity. And that makes his endorsement of Hapgood's original book on the phenomenon very interesting, to say the least. Not only was Hapgood well-connected academically, having come from Harvard, but let's not forget his OSS/CIA connections. It seems to me that if the big brains at the Pentagon were concerned about a sudden physical polar axis shift and the public's reaction to it, then putting out the information exactly as Hapgood and Einstein did is exactly the way they would want it done. If nothing else, it would spur geologists and other scientists outside the

Pentagon, who for whatever reason couldn't be brought inside to study the issue, to start looking into it seriously. And it certainly does create an interesting speculative time line:

1948: The Pentagon concludes a sudden spin axis shift will take place in the next 17–1,000 years.

1950: Emmanuel Velikovsky publishes *World's in Collision*, which declares that a nearby pass of large comet (which he says is now Venus) has tipped the Earth on its axis and back again.

1957: President Eisenhower declares "The International Geophysical Year," to study, among other things, plate tectonics, the rotation of the Earth, the magnetic polls, and so forth.

1958: Hapgood publishes *The Earth's Shifting Crust* with Einstein's endorsement.

1961: President John F. Kennedy sets the national goal of sending a man to the Moon and returning him safely to the Earth by 1970.

Mid-1960s: Rand Corporation publishes Habitable Planets for Man, a report which concludes that if the Earth's magnetic and physical poles coincide, the crust of the Earth will tear from the mantle and slide over by 90 degrees.

1970: Hapgood publishes *The Path of the Pole* in which he backs off the cataclysmic pole shift scenario in favor of a more gradual slippage of the physical poles because he doesn't have a plausible mechanism for sudden shifts. Plate tectonics theory generally accepted as an explanation for the erratic shifts in magnetic poles.

Mid-1970s: Mayan calendar decoded, 2012 date established and pole shift scenarios connected to it.

1976: The 12th Planet is published by Sitchin, providing a causal mechanism for the crustal displacement theory.

1985: First studies on true polar wander are published, concluding that the physical poles have wandered from their present positions, but only by about 1 degree every million years.

2002: Further studies show that there have been two dramatic 55-degree shifts in the spin axis of the Earth, but they occurred more than 700 million years ago and were probably due to comet or asteroid impacts.

As I add all this up, I see a pretty clear pattern. The shifting magnetic poles caused a great deal of alarm in the inner circles of government right after World War II. Scientists on the inside became convinced that the location of the Earth's spin axis was erratic and could flip on at almost any time. In an effort to get more information without alarming the public too much, several speculative works were published to stimulate scientific interest. Perhaps in response to the information and predictions, President Kennedy pushes for a manned space program to prepare an escape plan. In the mid-1960s, studies by the Rand Corporation conclude that, if the magnetic and physical poles meet, the crust of the Earth will break free and slide by as much as 90 degrees. By the 1970s, plate tectonics and how they affect the history of the magnetic pole records show that the crust is not likely to break free in once piece under any conditions. In the mid-1970s, publication of Sitchin's *The 12th Planet* and the decoding of the Dresden Codex lead to a renewed concern about the possibility of an axial pole shift. By the 2000s, new studies show that there have not been any major spin axis shifts for hundreds of millions of years at least, and those that are theorized were probably caused by asteroid or comet impacts.

• • •

So all in all, I'm far less concerned about the notion of a *2012* movie-style pole shift event than I have ever been. I don't think for a moment that the 2012 era will see such an occurrence. But I am still concerned about solar flares and maybe even the effects on our weather of a magnetic pole shift, at least to some degree. However, there is one thing that both the Z-pinch solar flare and the magnetic pole slippage concept have in common; it's called HAARP.

The High frequency Active Auroral Reasearch Program, commonly known as HAARP, is a very specialized ionospheric research facility located near Gakona, Alaska. Supposedly set up to experiment with ways to enhance naval communications and possibly to be used to map underground facilities in other countries like North Korea, HAARP has always been the subject of extensive speculation from conspiracy theorists as to its real purpose. These ideas range from mind control to earthquake creation. I have always felt it had another secretive and perhaps even dual purpose.

What HAARP does in essence is "harden" the ionosphere, the layer of ionized plasma that exists between the upper layers of Earth's atmosphere and the magnetosphere, the protective magnetic field of the Earth. Although the magnetic field does a fine job of protecting us from normal levels of harmful solar radiation like x-rays and other types of radioactive particles, it is weakest at the poles, and as we have learned that is where a huge solar flare could leak through the protective layers and scorch the Earth. Because the ionosphere is charged by the solar energy that leaks through the magnetic field, it essentially acts as a last line of defense from such harmful solar flares. By "enhancing" the ionosphere, which is what

HAARP's public literature admits it does, HAARP has the theoretical capability of adding an extra layer of protection in a Z-pinch scenario.

According to the published fact sheets on the project, HAARP is supposed to operate in the 3.6 megawatt range. By contrast, a typical terrestrial radio station in a major city operates at a mere 50,000 watts. Now, although 3.6 megawatts may seem like a lot, it isn't really enough energy to make a significant impact on the ionosphere. However, HAARP has more than 180 operating antennas, and combined together they yield an energetic output in the 5.1 terawatt (trillions of watts) range. As a comparison, the total energy usage of the entire human race for 2006 was about 16 terawatts.

At this level, HAARP could theoretically increase the strength of the ionosphere to the point that North America could be protected in the event of a Z-pinch solar flare. It could also, theoretically at least, be used to keep the magnetic poles from slipping too close to the geophysical pole and causing the crust ripping scenario the Rand Corporation was so concerned about in the 1960s. Whether or not is actually being used for this purpose would be pure conjecture on my part. In any event, that possibility would seem to be of minimal concern now anyway, given the more recent studies that would seem to refute the earlier Rand Corporation reports.

Other such facilities have sprung up elsewhere in recent years. There is one called the EISCAT ionospheric heating facility near Tromsø, Norway, funded by the European Union, and the Russians have built the Sura Ionospheric Heating Facility near Novgorod, Russia. The United States also maintains a second HAARP-like facility near the Arecibo Radio Observatory in Puerto Rico. It doesn't take much to look at a map and realize that if the true purpose of HAARP is to protect against an incursion from a Z-Pinch type solar flare, these locations would have the effect of protecting the United States, Western Europe, and Russia from the effects of such an occurrence. The Arecibo facility would effectively protect the southern flank of the United States from anything creeping up from the south Polar Regions. So the truth be told, though I think all of these end of the world scenarios are worrisome, it is evident that our governments at least seem to be taking some precautions against them.

But the reality is that, although all of these technological solutions are comforting, according to the Hopi at least, technological prowess didn't save the last World of Man from catastrophic destruction. But remember: That was a World where mankind had completely lost its way, lost its true connection to God—a World where the people had "*turned away from natural laws and pursued only material things and finally only gambled while they ridiculed spiritual principles.*" We are far from that today. Yes, we have issues, and there are a great deal of materialism and spiritual indifference across the world today, but there is also light. The light of a new dawn. All we have to do is Choose it.

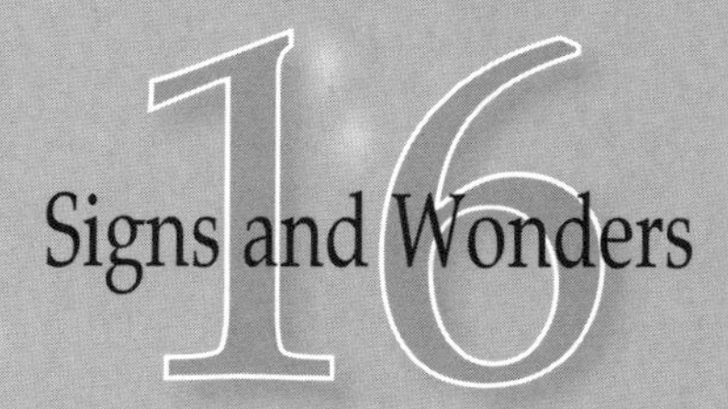

16 Signs and Wonders

For thousands of years, men have looked to the skies as the source of all knowledge, wisdom, and guidance. They have used the paths of the stars to track cycles and seasons, ages and epochs. Even today, in our arrogant belief that all is knowable and few things are mysterious anymore, we pay little attention to the planets and stars as they criss-cross the skies above us and the precession of the equinoxes ticks down to age after age. We are aware of these movements perhaps only subconsciously, rarely giving them much thought in our awakened state. But we don't have to be stargazers to feel the influence of the solar system and the galaxy all around us.

There was a time, back when we were in harmony with the Earth and in co-creation with God, when we understood this. A time before our minds became so active and powerful that we lost touch with the inner spirit that truly fuels our hearts and bodies here in the third dimension. So it stands to reason that if we simply look to the skies for signs and wonders as we once did, the next few years and all of their mysteries should be revealed to us.

The Maya certainly believed this. That's why they tracked the movements of Venus so precisely and then embedded them into the Dresden Codex. The Egyptians certainly believed this; that's why they built the Great Pyramids to track the movements of Orion's belt up and down the meridian from the vantage point of Giza. The Nazca Indians of Peru certainly believed it; that's why they inscribed the constellations above on the vast plains around their cities as an earthbound cosmic mirror. It is just our generation, with our left-brain-dominated modern perspective, that has plugged our ears to the songs of the stars.

As Professor Robert Temple, Fellow of the British Royal Astronomical Society, once put it in *The Sirius Mystery*:

> In my opinion, a mind is healthy when it can perform symbolic acts within mental frameworks which are not immediately obvious. A mind is diseased when it no longer comprehends this kind of linkage and refuses to acknowledge any basis for such symbolic thinking. The 20th century specializes in producing diseased minds of the type I refer to—minds which uniquely combine ignorance with arrogance. The 20th century's hard core hyper rationalist would deride a theory of correspondences in daily life and ritual as primitive superstition. However, the rationalist's comment is not one upon symbolic thinking but upon himself, acting as a label to define him as one of the walking dead.

Now, I wouldn't be quite as harsh as Professor Temple. After all, we are essentially brainwashed from childhood to listen only to our rational minds and ignore the songs of the soul. But he does make a valid point: Any mind that looks only at the rational and refuses to integrate the wonder of the emotional experiences in our hearts is truly "dis-eased." As I've joked many times, I am a card-carrying member of the military/industrial complex. In all my years as in the technical fields, I have not known many happy scientists or engineers. But everyone I know who spends more time in their hearts than they do their heads seems to lead a gloriously happy life.

The ancients understood this. That's why the scientist and the mystic were one until very recently in human history. They may not have been smarter than us in the literal sense, but they certainly got more out of their shorter life spans than most of us do. And all because they simply watched the stars...and wondered.

So as we feel our own spirits lifted, as the stars tug not on our conscious minds so much as our hearts, what will we see if we start to look up? As I look forward, I (like the Maya and the Hopi long before me) feel change in the air. We all want to know what will happen in 2012, and the truth is that what will happen is *already* happening. The number of unusual astronomical events taking place in the next couple of years (all of which were known by the Maya centuries ago) firmly convinces me that something is going to radically transform our world very soon.

Two critical celestial events will have already taken place by the time you read this. The first is the second Cosmic Convergence, which passed on July 17, 2010. Calleman sees this event as the first real chance to set the frequency of the planet before the possibly more chaotic upheavals begin. As he sees it, the purpose of the Galactic Underworld, the phase of consciousness we are currently in, is designed to bring about balance. Balance between not just East and West, but between left brain and right brain. Just as our world is dominated by Western thinking, so is our consciousness dominated by the logical mind, or the left brain. As we come closer to the convergence which begins on July 17, more and more of us will be seeking truth and light, more and more of us will be asking questions, and more and more of us will want to find a way to help others.

Calleman calls this new state of being a "supramental cosmic consciousness," which will come to fruition because of the driving forces behind the Mayan calendar, forces that, I might add, he still has no physical explanation for. As he put it:

> A crucial point of preparation for the attainment of this supramental cosmic consciousness will be the Second Harmonic Convergence/Oneness Celebration of the Gregorian date July

> 17, 2010. This is when for the first time, standing on the balance that the Galactic Underworld has brought, we will begin to sense the supramental consciousness that the Universal Underworld will bring. For all those seeking to become part of the new species of human beings endowed with an unlimited consciousness this is a time to absorb the new energies of universality. These dates are placed in relation to the Universal Underworld in the same way as the original Harmonic Convergence, August 16–17, 1987 might be seen to have been a preparation for the Galactic Underworld. July 17 is the date for the Phoenix to rise from the ashes and for an Enlightened world to arise from chaos. A sort of resurrection to the New Jerusalem where life is lived totally in the present moment with no limits and no separation. God is good. There is a logic to the evolution of the divine encoded by the Mayan calendar. Maybe, in fact, God, life, and consciousness are all the same.

So what many of us are experiencing right now, the awakening of friends and neighbors that we had never suspected would be interested in anything other than fashion magazines and reality TV, is being driven to new beginning in July of 2010. As Calleman sees it, the Conscious Convergence will result in a breakdown of monetary systems the world over, starting in the United States. He expects that sometime between July 17, 2010, and the end of the Sixth Night of the calendar on November 2, 2010, there will be a collapse of the Western banking system. He sees a possibility that all of the debts that are owed to banks will be simply forgiven, and that international debts will disappear in a similar manner. This might very well lead to war, but the last major transition of this sort was in the 1986–1992 period, and we had no wars to speak of then, even as the Soviet Union fell.

I don't see any of this happening, although I believe the Cosmic Convergence is a critical opportunity to focus the energy of our planet through prayer and meditation. The biggest flaw I see is that there is nothing in the physics that says that these dates are necessarily significant, or that there is anything special about the period. As we learned in the earlier chapters, alignments like the galactic core/Sun/Earth alignment or eclipses can have measurable effects on instruments and electronic devices, but only when these specific alignment scenarios occur. However, what Calleman may be seeing in the decoding of the calendar is the opening of a window in which the changing planetary positions (and therefore the physics of the Earth itself) are ripe for a consciousness change. We and our planet are becoming energetically pliable, like a huge glob of cosmic silly putty. If we look beyond the Sixth Night, we see a series of alignments and events that are virtually guaranteed to affect us in significant and unpredictable ways.

The first interesting celestial event takes place right after the symbolic end of the Sixth Night of the Mayan calendar on November 2, 2010. Just about six weeks later, as we are on the rise up the curve on the morning of the seventh day, there is a total lunar eclipse. That in of itself is not so unusual. Total lunar eclipses occur once every couple of years somewhere in the world. What is unusual about the 2010 total lunar eclipse is that it takes place on very significant date: December 21, 2010, the 2010 winter solstice. Two years to the day before the end of the current Mayan Long Count calendar cycle.

So just a few weeks after this book is printed, the Moon will turn blood red in the night sky over all of the United States and North America. As we've already seen from Maurice Allais's experiments, solar eclipses can have a profound effect on physical instruments here on Earth. What affect will this lunar eclipse have on a populace already in a heightened state of awareness due to the effects of the second Harmonic Convergence in May and the Conscious Convergence in July? Will it mean that we begin a new phase in the acceleration of the change we are already experiencing? It's certainly possible. But there is no question in my mind that the event itself is significant and that it will have an impact on our thinking to some extent. It's possible that what's conceived in May and birthed in July will be consecrated in some fashion in December. But there is just no way that an eclipse on the winter solstice exactly two years to the day before the end of the Long Count Cycle can be an accidental arrangement of celestial bodies.

May 20, 2012 Solar Eclipse

One of the reasons I think Calleman is wrong about his October 28, 2011 calendar end date is that there is nothing at all unusual astronomically in 2011. Yes, there are four partial solar eclipses and two total lunar eclipses, but I predict their physical/emotional effects will be minimal. However, the next major eclipse event takes place on May 20, 2012, and it I suspect that one will be a really big deal.

Eclipses always take place on a New Moon. Astrologically, New Moons are symbolic of new beginnings and new opportunities. So it stands to reason that our consciousness will be affected during this period because the eclipse is likely to amplify the already active tendency to look for a fresh start. In fact, as we look at the physics of this eclipse, it is almost a guarantee that we will experience a major shift in our thinking (if not our living conditions) around this time.

How this all relates to the May 20, 2012 eclipse is significant.

On that day, the annular solar eclipse will pass across the Northern hemisphere literally between East and West. (An annular eclipse means that the Moon and Sun are in perfect alignment, but the Sun is not totally blotted out because the Moon is a little too close to Earth to create a total solar eclipse.) The eclipse first begins (think of it as the first contact, similar to the external contact in the Venus transits) just off the coast of China, right on the Tropic of Cancer. The exact moment of greatest eclipse is literally on the International Date Line. The exact midpoint between East and West.

Kind of a neat trick, considering Calleman's take on the Galactic Underworld, isn't it? "A balance between East and West." That is literally what we are seeing. And, as you track the path of the eclipse, it gets even more interesting. By the time it reaches the shores of the United States, where the eclipse ends, the Moon is positioned directly over a very significant spiritual center: Mount Shasta, California.

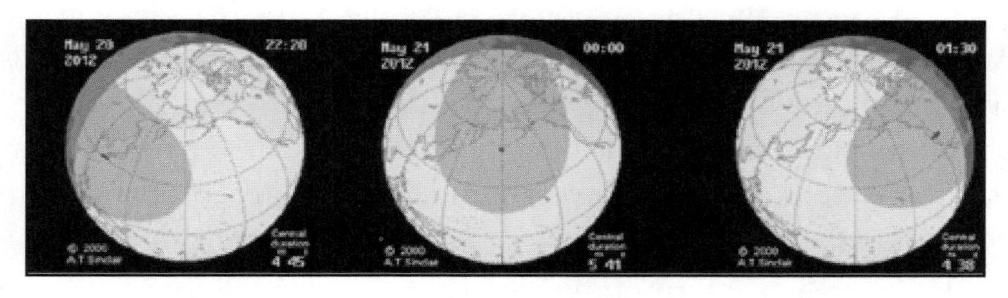

Image 16.1. The path of the May 20, 2012 annular solar eclipse coincident with a new moon. Image courtesy of NASA.

Mount Shasta has long been considered a sacred spiritual locus by many faiths. The mountain itself is completely isolated, towering more than 10,000 feet above the surrounding terrain. Local Indian tribes have long worshipped in the area, believing it to be hallowed ground that is used for spiritual ceremonies all year long. In addition, there is a Buddhist monastery on the mountain, and some New Agers believe that there is an underground base inhabited by the Lemurians, a sister race to the Atlanteans who came to the mountain after the destruction of their continent in ancient times.

Whatever you think of all that, the bottom line is that everybody who has visited the place believes it is sacred and special. But what will happen on the day of the eclipse is perhaps even more sacred and special. When the eclipse of May 20th comes to an end, the Moon's shadow will effectively be in the fourth contact position, identical to the external contact position of the Venus transit of 2004. If form holds true (and according to the physics it *must*) there will most likely be a profound and easily detectable effect on all kinds of electrically charged instruments, including, I suspect, our *brains*.

To me, being open to these kinds of resonances in the physics, this is amazing. To have an eclipse during a period of time that is supposed to be a uniting of East and West (according to Calleman) start exactly in the East (on the Tropic of Cancer) and reach its midpoint at exactly the International Date Line and end exactly in the West over the most sacred natural landmass in the United States is way beyond anything I can call coincidence. The point is: None of these coincidences *are* coincidences. They are all driven by the *physics*. Someone, somewhere *has set this all up*, eons ago, probably from the moment of creation itself.

If Calleman's interpretations are correct, this eclipse is an overwhelmingly important marker in the transition we are experiencing. It is literally the Mayan calendar come to life, expressed through the movements of the celestial bodies in the physical realm. This eclipse will bring balance, exactly as Calleman predicts the Galactic Underworld will do. What we are seeing in this eclipse is the physical manifestation of Calleman's vision of the Galactic Underworld. Balance between East and West. Balance between the North Pole and the equator. Balance between the masculine and feminine. Balance between the mind and the spirit.

Now, it can of course be argued that Calleman is flat wrong, because the transition of consciousness he describes takes place in the 2010–2011 time frame, not 2012. I am nowhere near an accomplished enough Mayan scholar to question his methods here, but what I will say is that in reading the cosmic tea leaves in the only way

I know how, through the movements of the planets, it seems to me that the transition he sees will be in 2012, not 2011.

That is not to say he is wrong. In fact, without question, I now concur that the Seventh Day will bring about a change in consciousness exactly as he describes. So Calleman, I think, is absolutely right. He's just got the dates a little mixed up.

But the key thing to remember, and what I worry about, is what happens next. According to Calleman's reading of the Mayan calendar, once we achieve this level of consciousness, everything is hunky-dory, sweetness and light. But my knowledge of the physics tells me otherwise, because after May 20, 2012, there are still three very significant astronomical events taking place before the Long Count cycle ends. And the changes they bring may have nothing to do with our thoughts and feelings, and everything to do with the physical state of the planet we live on.

Just a few days after the eclipse comes the June 6, 2012 transit of Venus, the second in the cycle that began in 2004. Then in November comes a second eclipse covering most of the Southern Hemisphere. And then, on December 21, 2012, we have the Yod and the other astrological alignments. What we have to do between now and then is to pay attention to these signs, to take the warnings and the portents seriously, and to use our own power, the power of spiritual thought, to set the intent of the planet for the future. We have, each of us, enough energy to make this world into anything we wish it to be. When Kukulkan arrives on his serpent rope to judge us, it is not the biblical judgment of Judgment Day we will face. It is his judgment of our intent. That is what the Hall of Mirrors is for. It is for us to Choose.

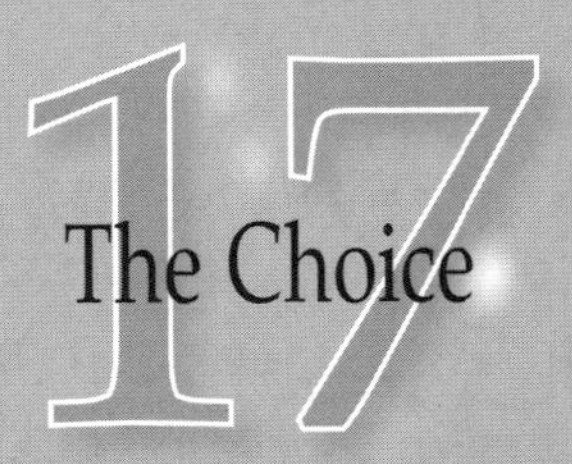

The Choice

So by now, your head is probably filed with so much information and so many data points that you aren't even sure what it all means. You probably also want me to get to the point: What do I think is going to happen in 2012?

What I think is that we're all going to have to make a Choice.

It's clear from all of the information that's out there that the 2012 era will be one of great change. The physics and the astrology say we are going to become even more polarized.

We are here, all of us, in this time, to take sides.

And by the way, it's too late to shirk your responsibility. By even picking up this book (hopefully you bought it!) you have already voted. You have decided to be one of the awakened ones. You have chosen to participate. What you are here to do is make a difference, not just for yourself, but for others. And, keep in mind that when you go down this path, there will be no glory, no reward. You will get no credit for your selfless participation in saving the planet. All you will get is ridicule from the sleeping ones when they lump you in with all the doomsayers and nothing really bad happens. It sucks, but it's what each of us has Chosen.

Keep in mind that one or even several of the physical catastrophe scenarios could still happen. It does appear that our governments are aware of that possibility, and seem to be taking some steps at least to safeguard our world. But it is also clear to me, by the haphazard way they seem to be preparing for *every* eventuality, that they are trying to predict or manage what is going to happen to whatever degree they can. What they can't predict, however, is just what our level of participation in our own future will be.

See, we are the wildcard. We have been taught, brainwashed even, to believe that the apocalypse, the time of sweeping change, is something that happens *to* us, rather than something we actively participate in. That is where I believe so much of the 2012 thinking has gone awry. It's not this specific scenario or that kind of disaster we have to worry about, it's our own Choice in the little matter of our own future. Many of the best 2012 researchers, even the ones I like and admire like David Wilcock and Daniel Pinchbeck, act as if Ascension is a done deal. As if all we have to do is show up and the world will be hunky-dory and we'll all spend the next 10,000 years singing Kumbaya. I don't agree.

Although it's true that the physics and alignments of the planets make that a distinct possibility, so too are some of the catastrophe scenarios I outlined. The stars cannot *compel*; they can only impel. Governments are trying to help. They are doing their best to create technological solutions to our problems, but that is only half the battle for the future. The other half must come from us.

Make no mistake about it: There will be a pulse of energy, torsion energy, from the center of the galaxy sometime in the next decade or so. The Mayan god Kukulkan will ride his "serpent rope" from the center of the galaxy to judge us. Kali will return to the Earth and slay the wicked and unfaithful. The Hopi Great White Brother will return to the sacred lands and judge our works, and sweep aside that which no longer serves God's higher purpose. Osiris will rise again to weigh each man's heart against the weight of a feather. But again, the mistake we all make is thinking of these judgments in terms of the Western, biblical Revelations model. You're good. You're bad. You go to Heaven. You go to Hell. It's not that kind of judgment.

The pulse of energy itself is neutral. It has no inherent intent until it manifests here in three-dimensional reality. What Kukulkan or the Great White Brother or Kali or Osiris will come here to judge is not his or her vision of who we should be, but our own *intent* for ourselves and our planet. If we are a world mired in fear, scarcity, greed, conflict, and selfishness, then the judgment we get will reflect that selfish spiritual orientation. But if we are a world alight in love, abundance, compassion, forgiveness, and selflessness, we will get a judgment that will perhaps allow us to see a new way, to participate in a true co-creation with God. To touch our Maker in a way that we have long forgotten, and that we still yearn for.

And no, we don't have to "earn" it. But we do have to Choose it.

God loves us enough to allow us our free will, and he does not hold it against us when we fail to Choose him. It is we who suffer the karmic price when we turn away from Him. God does not suffer. We are the ones who cry and hurt and bleed for our sins. And he does not have to forgive us. He never held our Choices against us. It is we who have to, as the first step, forgive ourselves. As the Mayan prophecies state, we will all have to face our personal Hall of Mirrors, where we must confront our own Choices and conduct and, I suspect, pass judgment on ourselves before Kukulkan passes judgment on the intent of our planet.

So how do we do this? If it is true that we get to participate in the End of the World as we know it, how do we do so? How do we guarantee that we get to be in the magical 144,000 to ask the right question, and in the right way? That's easy. All we have to do is ask ourselves.

To my mind, this comes down to only two very simple questions: Who are you? And what do you want?

One of the great travesties of our time is the misinformation being given to us today by so many of the so-called spiritual gurus around the world. Popularized by such tripe as The Secret, they will tell you that all you have to do to have a perfect life is to ask for it. But what they sell you is a perfect life of material things. The Secret works great if all you want in life is a big house, a fast car, and hot girlfriend. But what if that isn't really who you are? And what if, at a spiritual level, it isn't really what you want? It is not enough to get down on your knees and pray for *things*. I admit, I do it. I want a Porsche in my driveway. But I also set my true *intent* before I ask Great Spirit for the gift. And in order to do that, I have to be honest with myself about *who I am*.

If I want the Porsche just to show it off to others and show them that somehow I am better than them, I probably won't get it, at least not in the way I want it or in the time frame I want it. But if I truly want it because I just love how it looks, how it responds to my steering inputs, and how much fun it is to drive, then I am doing it for my self-interest, not my selfishness. Now, that doesn't mean that my intent has to be 100 percent pure. Yeah, I'd love to take a Porsche or Bentley or a Ferrari to cruise around Hollywood with. Why not? It would feed my ego, and that is an aspect of my personality, just as my personality is an aspect of who I am. But if my intent is to use it to put others down, to make them feel less valuable so that I can feel more valuable, then why would Great Spirit or the Universe give it to me?

Maybe to teach me a lesson. Maybe that's why bad people sometimes end up rich or famous or seem to get away with doing horrible things. But I don't think that happens in 99 percent of our lives. What I think is that the vast, vast majority of us have come here to learn to be good. To leave this life closer to the Maker than when we arrived. And in the course of that journey, if I decided, on a spirit level, that I had to *earn* my fortune, then no matter how much

I prayed to God or Great Spirit or the Universe to win the lottery, it isn't going to happen. That would be at cross purposes with my life lesson. So my first task is to learn who I am before I can set about the task of following the path I agreed to before I came here.

My maternal grandmother, Hazel, once said to me, "If you always tell the truth, you'll never have to remember what you said." Like most of us, I lie all the time. Mostly little white lies, but still lies. I lie to make myself feel better. I lie to keep from hurting the feelings of others. I lie sometimes to manipulate others to get what I want from them. We all do. But if that is the biggest part of who we are, then that will be the biggest part of what we reap. So the first step is to not lie to yourself about who you really are.

How do we do this? How do we even know what our life purpose is?

Not to give you a spiritual platitude here, but the answer lies within. Find something that works for you, something that touches your heart and activates your connection to God, for that is the place you will find out who you truly are in this life. I don't care how you do it. For some, it will be the Mormon Church, or the Catholic Church, or the Kabbalah or Yoga, or the Self Realization Fellowship or the Aetherius Society. But find something that works for you.

For me, my spiritual center comes through music, mostly. I love the soaring melodies and poetic lyrics of U2. No matter how many times I've heard the haunting desperation of "With or Without You," or soaring joy of "Beautiful Day," the dark lament of One, or even the quiet loneliness of "Moment of Surrender," I am moved. Trance music and New Age Music like Enya or the work of Steven Halpern can put me into my heart very quickly. But prayer and meditation? Chanting? Not so much. The point is to try lots of things. Find what works for you, and take time to find that place every day.

But keep in mind that simply going to that place isn't enough. Yes, it can calm us when we are anxious, soothe us when we are sad, fill us when we are empty. But that, like The Secret, is all about *us*. What do we do once we get there? Meditation is fine, but the purpose of meditating is to empty our minds to get in touch with our spirit. But then what? Imagine God or Great Spirit sitting there tapping his toes and saying "Okay, nice to talk to ya again. Yeah, Mike, I'm working on that Porsche. And yeah, I'm working on the girlfriend. Now what?" It is at that point that you can set your intent to the Universe for the kind of world you want to live in. It is at that point that you can get Kukulkan to pay attention to your vote, and to please drop the whole solar flare/pole shift idea and instead send some energy to alleviate the suffering of your brothers and sisters.

One church I like to attend once in a while is the Agape International Spiritual Center in Culver City, California. It is presided over by Michael Bernard Beckwith, a preacher in the charismatic traditions of black churches throughout the South. I like to go there because, though it is a very mixed group racially, the church also has a very black feel to it. I like that. It makes me a little uncomfortable because I'm so damn white, for one thing. But it's okay to be uncomfortable once in a while. It gives you a chance to learn something about yourself. (Just don't expect me to keep time with the music.) But one thing I notice about Agape is that it really isn't any different than any other traditional churches I have attended. MBB packs 'em in for three services on Sundays and again on Wednesdays, primarily because he was a participant in *The Secret* documentary film and that got him on *Oprah*. But what I see is a couple thousand people at each service. They come in. They sing. They listen. They laugh. They cry. They hold hands. And then they leave and don't really come back to that place in their hearts and minds again until the next week's services. They don't seem to realize that God is not a vending machine. God is something you have to live and breathe every day, and take with you out into the world. It is by doing that, by looking for ways to help others, by responding to honest requests for assistance, that you will find out just how much giving there really is in you. You will come to know yourself. And once you know that, you can answer the second question: What do you want?

If you want a world without war, don't protest against war. Rally for peace. Sometimes wars are necessary. They are! World War II was a prime example. There is true evil in this world. And yes, though it's true that if God created goodness he must also have created evil, it is still evil. It still intends to harm you and those you love. It still intends to deceive you with material pleasures, to tempt you into turning away from God. Evil wants us to make the same Choice that the other three Worlds the Hopi speak of did. It wants us to stay down here in the netherworlds with it, in these lower vibrations, because it is lonely and desperate and vile down there. So sometimes you have to oppose evil, no matter how much you dislike war. God won't do it for you. He can't. Has his own Prime Directive. Like on *Star Trek*, he can't intervene. We have to. We have to Choose good over evil.

Maybe you want a world without money. Maybe you reject the whole moral underpinnings of the commercial system. But look around you. It has served us well. Look at our hospitals and our miracle drugs and our space stations. Look how far we've come since Jesus came and offered us the ultimate New Deal. Most of it was built on money and credit. Is it really such a bad system after all?

So think about what you want, truly and deeply and in your heart and soul, for the world around you. If you don't like commerce, don't ask for something different; ask for a world of greater abundance and generosity, with less suffering and material envy. Let the Universe decide how to give it to you.

But keep in mind that *something* is going to happen in the next few years. Something profound and different. The planets that are driving the physics demand it. Put simply, there are far too many astronomical coincidences in 2010 to 2012 for nothing at all to happen.

We have several key pressure points in the system leading up to 2012. By the time you read this, several of them will have already passed. But there is still time. The sixth night of the Mayan Calendar ends (and the seventh day begins) on November 2, 2010. Just a bit after that, we have the winter solstice eclipse of 2010, just two years to the day before the Long Count runs out. Then we have the end of the Aztec calendar cycle on October 11, 2011. Then the 2011 winter solstice, the May 2012 Mt. Shasta eclipse, the 2012 Venus transit, and the November 2012 eclipse occur in rapid succession. Finally, the Long Count ends on December 21, 2012. By that time, I believe we will have set the intent of the planet for the next Age. Kukulkan, when he arrives here, whether it is on that day or some day in the near future, will get the message. We just have to decide. You have to decide.

So who are you? And what do you want?

I can't wait to hear what your answer is.

Chapter Notes

Chapter 2

1. Dwarf planets are defined by the International Astronomical Union as objects that freely orbit the Sun—that is, not moons, but that are not massive enough to be rounded by their own gravity.

Chapter 3

1. Interestingly, the simplest solid geometric form known to us is the tetrahedron, and the most complex is the sphere.
2. The sunspot activity of the Sun follows a pattern of rising and falling over a 22-year-cycle which governs the Sun's magnetic field. Every 11 years, the activity reaches a peak and magnetic poles flip, after which the sunspots quiet down before cycling up again for the next rising cycle.
3. *www.time.com/time/magazine/article/0,9171,814720,00.html*
4. "Shortwave Radio Propagation Correlation with Planetary Positions." RCA Review 12 (March 1951): 26–34.
5. In a famous sequence in TV series *Cosmos*, Carl Sagan used this analogy to dismiss astrology: "Gravity can't be the cause, because while the planets are much more massive, the doctor who delivered you was far closer."
6. *Dark Mission-The Secret History of NASA* Chapter 2, pp.22, 23. Feral House, 2007, ISBN: 978-1-932595-26-0.

7. "RCA Astrology." Time magazine, Monday, April 16, 1951 *www.publish.csiro.au/rid/138/paper/AS06018.htm*
8. *Does a Spin–Orbit Coupling Between the Sun and the Jovian Planets Govern the Solar Cycle?*
9. *"RCA Astrology."* Time magazine, Science: Monday, April 16, 1951.
10. The other condition Nelson mentioned, the 0° configuration, is called a conjunction, and occurs when a two or more planets all appear essentially in the same place, right on top of each other. In Nelson's studies, this tended to have a tense effect on the Sun, causing magnetic storms much like the square and opposition. Although conjunctions are considered neutral in astrology, they can take on aspects of the planets involved. For instance, I don't know if Nelson noted how many times Mars, the planet of conflict, was involved in the conjunctions he studied.

Chapter 4

1. There is actually a debate over whether light is made up of particles or waves, but that's another book.
2. *www.nature.com/news/2008/080813/full/news.2008.1038.html*
3. Sub-atomic simply means smaller than an atom, which is so small that no one has ever seen one. We can be pretty sure that they exist though, because some pretty big stuff happens when we split one.
4. The word *atom* comes from the Greek *atomos,* by the way, which means "indivisible"—oops.
5. The word *Baryon* comes from the Greek *Bayrs*, meaning "heavy."

Chapter 5

1. The Julian calendar had been adopted by the Christian Church at the council of Nicaea in AD 325, the first of the general councils of the Church.
2. Proceedings of the American Antiquarian Society at the semi-annual meeting in Boston, April 8, 1914, *Notes on the calendar and the almanac*, George Emery Littlefield, p.p. 29

3. The meridian is the imaginary north/south line that passes directly overhead from any given vantage point on Earth. When the Sun hits high noon, it is dead on the meridian.
4. *The Death of Gods in Ancient Egypt,* Penguin, London, *1992.*

Chapter 6

1. Portions of this section are reproduced from *www.salagram.net/cycleOages.html.*
2. Sri Yukteswar, Swami (1949). *The Holy Science*. Yogoda Satsanga Society of India.
3. Ibid.

Chapter 7

1. A "light year" is the distance that light travels in a vacuum in one year, approximately 6 trillion miles (5,878,630,000,000). As a comparison, Neptune, the most distant of the planets is 0.000963 lightyears away, about 4.16 light hours.
2. Raup, D.M.; Sepkoski, J.J. (1984-02-01). "Periodicity of Extinctions in the Geologic Past." *Proceedings of the National Academy of Sciences* 81 (3): 801–805. doi:10.1073/pnas.81.3.801

Chapter 8

1. Solstices are two of the four astronomically significant dates in the year that have to do with the Earth's motion around the Sun. The winter and summer solstices (December 21st and June 21st) are the shortest and longest days of the year, respectively, and the spring and fall "equinoxes" are the dates when the night and day are of equal length.

Chapter 11

1. *www-istp.gsfc.nasa.gov/Education/wtether.html.*
2. Dr. Noble Stone, TSS-1R NASA Mission Scientist *www.enterprisemission.com/tmm0010.html.*

Chapter 12

1. "... we all realized immediately that the rocket had provided a larger than expected thrust, resulting in a higher than planned orbit, and a longer orbital period. The orbit had been expected to have a perigee (lowest height above the Earth) of about 220 miles and an apogee (greatest height) of about 1000 miles. The perigee and apogee heights were actually 223 miles and, more significantly, 1592 miles, respectively, with an orbital period of 114.7 minutes rather than the 105 minutes that had been originally anticipated." *http://www-pw.physics.uiowa.edu/van90/ExplorerSatellites_LudwigOct2004.pdf.*
2. *www.enterprisemission.com/Von_Braun.htm.*
3. Allais, Maurice, "The Allais Effect and my Experiments with the Paraconical Pendulum 1954–1960" Report for NASA, 1999 *www.allais.info/alltrans/nasarep.htm.*
4. *http://home.t01.itscom.net/allais/blackprior/noever/decrypting.htm*
5. *Pushing Gravity: New Perspectives on Le Sage's Theory of* Gravitation – Matthew R. Edwards ISBN: 978-0968368978.
6. *www.allais.info/priorartdocs/noever.htm.*
7. *http://astro.univie.ac.at/.*

Chapter 13

1. *http://www.uwgb.edu/dutchs/platetec/rotationqk2004.htm*

Chapter 15

1. Gold, T. "Apollo II Observations of a Remarkable Glazing Phenomenon on the Lunar Surface." *Science* 165 (1969): 1345.
2. Contrary to what you may see reported on the news, the Sun, not human industrial activity, accounts for climate change and temperature fluctuation here on Earth.
3. Dumé, Belle. March 21, 2006. "Geomagnetic flip may not be random after all." *physicsworld.com*. *http://physicsworld.com/cws/article/news/24464*. Retrieved December 27, 2009.
4. Ken White. *World in Peril*. page 197.

Selected Bibliography

Allais, Maurice. "The Allais Effect and my Experiments with the Paraconical Pendulum 1954–1960." NASA, 1999.

Bara, Mike, and Richard C. Hoagland. "Chapter 2." In *Dark Mission—The Secret History of NASA*. Feral House: 2007.

Dumé, Belle. March 21, 2006. "Geomagnetic flip may not be random after all." *Physicsworld.com*. (2009).

Edwards, Matthew R. *Pushing Gravity: New Perspectives on Le Sage's Theory of Gravitation*. Apeiron, (2002).

George Emery Littlefield. "Proceedings of the American Antiquarian Society at the semi-annual meeting in Boston." (1914): *Notes on the calendar and the almanac*. 29.

Gold, T. "Apollo 11 Observations of a Remarkable Glazing Phenomenon on the Lunar Surface." *Science* 165 (1969):1345.

Nelson, J.H. (1951). Shortwave Radio Propagation Correlation with Planetary Positions. *RCA Review* March, 26–34.

Raup, D.M.; Sepkoski, J.J. "Periodicity of Extinctions in the Geologic Past." *Proceedings of the National Academy of Sciences* 81 (1984): (3): 801–805.

Salart, D., A. Baas, C. Branciard, N. Gisin, and H. Zbinden. "Quantum mechanics: The speed of instantly." *Nature*, 454, (2008): 861–864.

Sellers, Jane. *The Death of Gods in Ancient Egypt*. London: Penguin, 1992.

Sri Yukteswar, Swami. *The Holy Science.* Yogoda Satsanga Society of India, (1949).

White, Ken. *World in Peril.* K. W. White & Associates: 1994.

Wilson, R.G. B. D. Carter B and I.A. Waite. "Does a Spin–Orbit Coupling Between the Sun and the Jovian Planets Govern the Solar Cycle?" *Publications of the Astronomical Society of Australia.* 25(2) (2007): 85–93.

Index

Accretion model, 37
Aether, 16, 54, 134, 135, 136, 153, 161
Age of Aquarius, 77, 189
Age of Pisces, 77, 189
Ages, 81-90
Aldrin, Buzz, 191
Allais Effect, 150-154, 161, 164, 169
Allais, Maurice, 150-154, 158, 170, 213
American Revolution, 160, 161, 165
Apollo 11, 191
Arguelles, Jose, 113-114
Armstrong, Neil, 191
Astrology, 31, 54-58, 121, 214
Astronomical conditions, 155-158
Astronomical Society of Australia, 53
Astronomy, 31, 54-58, 121
Astronomy, the Maya and, 102
Backster, Cleve, 22-24, 64
Balance, 216
Ballistic trajectory, 145-146
Banyacya, Thomas, 123-127, 129
Baryonic matter, 66
Bauval, Robert, 78
Beckwith, Michael Bernard, 224
Biburu, 198-199
Bode's law, 39, 40, 42
Bolon Yookte, 115-116, 118-119
Bolon Yookte, 174-175
Book of the Hopi, The, 125-127, 129-130
Braden, Greg, 174
Brahma Vaivarta Purana, 98-99
Brahma, 82-90
Brain waves, 57
Bronze Age, 82-84, 85, 88, 92, 99
Brookings Report, 202
Brown dwarf, 181-182

Caesar, 73-74
Calendars, 72
Calleman, Carl Johann, 114-121, 161, 211, 214-216
Canis Major, 74
Caran, Eli, 135-138
Ceres, 40
Chandler wobble, 171-172
Cleopattra, 73
Consciousness, 10, 119, 121, 186-187
Cortex, Hernan, 103, 160
Cosmic Convergence, 211-214
Cotterell, Maurice, 107-108, 111
Course of Astrophysics and Stellar Astronomy, The, 155
Crustal displacement, 193-198
Dark matter, 66
Dark Matter, Missing Planets and New Comets, 36, 40
Dark Mission, 15, 16, 32, 43, 46, 77, 128, 161, 173, 178, 202
De Landa, Diego, 103
DePalma, Bruce, 61-62, 139, 143, 149, 158, 162
Dharma, 81-82, 87
Dimensions, 16-17
Direct Ascent, 147
Dissonance, 33
Divine deminine, 160-161
Djed pillar, 79, 116
Dreamspell Count, 114
Dresden Codex, 12, 13, 101-107, 108, 111, 114, 159, 174, 204, 209
Dwapara Yuga, 86, 88
Earth's Shifting Crust, The, 196, 197, 203
Edwards, Matthew R., 153
Einstein, Albert, 15, 16, 23, 59, 63-66, 69, 135, 150, 156, 196, 197, 200, 202-203
Einstein-Cartan theory, 138
Electricity, 50
Elohim, the, 19-24, 27
End of Days, 173
End of Time, 107
ESP, 96-97
Exploded Planet Hypothesis, 39
Explorer Effect, 143-150, 153-154
Explorer I, 143-150, 153
External contact, 163
Falkov, Robin, 200
Faraday cage, 139-140
Feynman, Richard, 65
Fictitious forces, 68-70
Flynn Effect, the, 28-29
Flynn, James, 28
Foucault pendulum, 151
Foucault, Leon, 151
Freemasonry, 79
Fulcanelli, 192
Galactic Alignment, 165-172

Galactic Equator, 167, 169
Galactic superwave, 191
Galactic Underworld, 211-214, 216
Geometry, 162
Gilber, Adrian, 107-108, 111
Giza plateau, 78
Global Consciousness Project, 24-29
God force, 15-29
Gold, Thomas, 191-192
Golden Age, 82-84, 92, 99
Golden Alignment, 98
Gotze, Johann Christian, 102
Gravitation, 64
Great Flood, 183-184, 187, 193, 196
Great Pyramid at Giza, 75, 113-114, 184, 209
Gregorian calendar, 72-73
Grolier Codex, 101
Gyroscopes, 149
Haab, the, 101
HAARP, 205-206
Hancock, Graham, 74-75, 78, 183
Hapgood, Charles H., 195-198, 200, 202-203
Haramein, Nassim, 135
Harmonic Convergence, 13, 113-114
Harmony, 33
Hawking, Steven, 57
Heaven's Mirror, 75
Heaviside, Oliver, 16-17, 69, 135
Heim, Burkhard, 154
Henry, William, 174
Hoagland, Richard C., 15, 32-33, 43, 46-46, 47, 144, 147, 161-172, 173, 175, 200
Holy Science, The, 86, 94
Hopi prophecies, 13, 122-132, 187, 191
Hunahpu, 111
Hutton, William, 171
Hynek, J. Allen, 180-182
Hyper-dimensional physics, 11-12, 45-58
Hyperspace, 47-48
Infrared Astronomy Satellite, 182
Internal contact, 163
International Space Station, 128
Intuition, 29
IQ, 28-29
Iron Age, 82-84, 85, 92, 99
Isis, 74, 78, 79, 128
Jenkins, John Major, 113, 118, 159-160, 161, 165-166, 167, 174
Julian calendar, 72-73
Kaku, Michiu, 57, 135
Kali Yuga, 86, 87-88, 89, 98-99, 106, 191
Kennedy, John F., 203, 204
Kirlian photography, 20-22

Kirlian, Semyon, 20-21
Kozyrev, Nikolai Aleksandrovich, 154-158
Krishna, 88
Kukulkan, 160, 174-175, 217, 220, 221
LaViolette, Paul, 191, 194
Laws of motion, 60-63
Laws of physics, 46, 51, 59-70
Local Sidereal Time, 97-98
Long Count calendar, 93, 105, 111, 159-160, 173, 176, 213, 217
Lunar Orbit Rendevous, 147-148, 150
Lying, 223
Macaw, Seven, 111, 174
Madrid Codex, 101
Magnetic pole shift, 193-195
Mahabharata, 88, 89
Maldek, 42, 43, 44
Massau'u, 124
Matuyama, Motonori, 194
Maxwell, James Clerk, 15-17, 47, 69, 133, 135
Maya Cosmogenesis 2012, 165-166
Mayan Calendar and the Transformation of Consciousness, The, 120
Mayan calendar, 10, 11, 13, 101-111, 113
Mayan Factor, The, 113-114
Mayan Prophecies, the, 107-111
Mayan Prophecies, The, 108
Mayans, 11
Message of the Sphinx, The, 78
Milky Way galaxy, 94, 97, 166, 174-175
Monuments of Mars, The, 45, 47
Morton, Sean David, 19, 120, 200
Motion, laws of, 60-63
Mount Shasta, 215
NASA, 141, 153, 180, 202
National Science Foundation, 31
Nebular hypothesis, 34-35
Nelson, John, 49-58, 153, 185
Nelson, Roger, 24
Nemesis Effect, 172, 173-188
Neugebauer, Gerry, 182
Newton, Isaac, 15, 59, 60-63, 65, 69, 133
1984, 9
Noever, David, 152-153
O'Leary, Brian, 23
O'Neill, Tip, 57
Olbers, Heinrich Wilhem
Matthaus, 40
Olmecs, 101
One-Hearted People, 131
Opposition, 55-56
Orion, 74, 209-210
Osiris, 74, 78, 79, 220, 221

Pahana, 126
Paraconical pendulum, 151-152
Paris Codex, 101
Path of the Pole, The, 197, 203
Pendulums, 151-153
Phaeton, 42, 43, 44
Physics of time, 71-80
Physics, hyper-dimensional, 11-12, 45-58
Physics, laws of, 46, 51, 59-70
Pinchbeck, Daniel, 220
Pioneer 4, 145
Pioneer missions, 177-180
Planet X, 175-188
Planetesimals, 35
Planets, birth of the, 23
Plate tectonics, 203
Plato, 175
Pleistocene epoch, 201
Popol Vuh, 106, 111
Precession, 75-76, 165
Princeton Global Consciousness Project, 164
Project Nanook, 200-201
Puthoff, Hal, 139
Quantum physics, 15, 60, 69, 133, 135
Radin, Dean, 25
Ranger missions, 146-149
Raup, Donald, 95
RCA, 49, 50, 51, 55, 153, 185
Reimann, Georg, 45
Relativity, 63-66
Resonance, 33
Rocket engines, 145
Rocket Equation, the, 144
Sagan, Carl, 63
Sagittarius A., 97
Satya Yuga, 84, 86
Sciama, Dennis William, 138-139, 143
Secret, The, 14, 222, 224
Sepkoski, Jr., J.J., 95
September 11, 2001, 25-28
Serpent rope, 174-175
Set, 78
Shiva, 82, 98
Silver Age, 82-84, 92, 99
Sirius Mystery, The, 19, 210
Sirius, 128
Sitchin, Zechariah, 19, 198-199, 204
Solar fission theory, 35-39
Solar system, 31-44
Solving the Greatest Mystery of Our Time, 120
Sosigines, 73
Space race, 143-144
Space-time, 63, 156
Special relativity, 63-66
Spottiswoode, James, 96-97
Sputnik I, 143

Sri Kalka, 99
Stalin, Joseph, 154
Stone, Noble, 141
String theory, 60
Sunspot cycles, 49
Super Earth, 43
Temple, Robert, 19, 210-211
Thermonuclear reactions, 157-158
Time, 156-158
Time, 54-55
Times, physics of, 71-80
Topologists, 47-48
Torsion, 133-141, 143-144
Torun, Erol, 47-48
Tree of Life, 116, 120
Treta Yuga, 84, 86
12th Planet, The, 204
Two-Hearted People, 130-132
2012, 10, 172, 195
Tzolkin, the, 101
Underworlds, 116-118
United Nations, 123-125, 129-130
Universal consciousness, 13
Van Flandern, Tom, 35-39, 41, 42, 52, 178-179
Vanguard missions, 143, 148
Velikovsky, 200
Velikovsky, Emmanuel, 196
Venus transits, 159-172
Vesas, 11
Vishnu Nabhi, 12, 81, 82, 93, 96, 98, 99, 168
Von Braun, Werner, 144-150, 153-154, 158
Waters, Frank. 125
Weidner, Jay, 192
White Feather, 128-129
White, Ken, 200-201
White, Maynard E., 200-201
Wilcock, David, 135, 220
World's in Collision, 196
World's in Peril, 200, 203
Xbalanque, 111
Xibalba, 116
Yaxche, 116
Yod, 185, 217
Young, David, 125
Yudhisthira, Maharaja, 86
Yuga cycles, 81-90, 91-99, 174
Yuktsewar Giri, Swami Sri, 86-88, 91-99
Zapotecs, 101
Zep Tepi, 184
Zero-point effect, 140
Z-Pinch solar flare, 190-193, 194-195, 205-206

About the Author

Mike Bara is a *New York Times* best-selling author, screenwriter, and lecturer. He began his writing career after spending more than 25 years as a "card-carrying member of the Military-Industrial complex" where he worked for a wide variety of aerospace companies as an engineering consultant and designer. Mike's first book, *Dark Mission-The Secret History of NASA* (co-authored with the venerable Richard C. Hoagland), was a *New York Times* best-seller in 2007. Mike was born and raised in Seattle, Washington, attended Seattle Pacific University before embarking on his engineering career, and has lived in Seattle; Toronto, Canada; Chicago; and Minneapolis, Minnesota. He has an identical twin brother, David, an older sister, and two cats, Aurora and Muffy. He currently resides in Redondo Beach, California.